Edible Wild Mushrooms
OF ILLINOIS AND SURROUNDING STATES

Edible Wild Mushrooms of Illinois
& Surrounding States

A FIELD-TO-KITCHEN GUIDE

Joe McFarland and
Gregory M. Mueller

UNIVERSITY OF ILLINOIS PRESS
URBANA AND CHICAGO

© 2009 by Joe McFarland and Gregory M. Mueller
All rights reserved   Printed in Korea by
Tara Printing through Four Colour Book Group
P 5 4 3
∞ This book is printed on acid-free paper.

Library of Congress Cataloging-in-Publication Data
McFarland, Joe, 1963–
Edible wild mushrooms of Illinois and surrounding
states : a field-to-kitchen guide / Joe McFarland
and Gregory M. Mueller.
p.   cm.
Includes index.
ISBN 978-0-252-07643-5 (pbk. : alk. paper)
1. Mushrooms, Edible—Illinois—Identification.
2. Cookery (Mushrooms)
I. Mueller, Gregory M. (Gregory Michael)
II. Title.
QK605.5.I3M34   2009
579.6'163209773—dc22   2008040179

To the late Eugene Scheuring,
who generously took his grandson hunting
for wild mushrooms just once.

Contents

Preface   xi

Acknowledgments   xv

Do Not Ignore This Warning!   xvii

1. Tips for Beginners   1
How to Find, Identify, and Understand Wild Mushrooms

Mushrooms grow everywhere. Here's how to start finding and identifying the best ones to eat—and how to bring them to the table.

2. Common and Poisonous   19
A Few Toxic Mushrooms

Poisonous mushrooms grow everywhere in Illinois. Before you pick and eat any wild mushroom, learn to recognize toxic species.

3. Into the Forest   33
Mushrooms Found with Trees

Edible mushrooms often grow on or around living and dead trees.

4 The Morels 79
Morchella esculenta, Morchella elata, and Morchella semilibera

The most popular wild mushrooms in Illinois look like a sea sponge. There are three basic morel species—plus several varieties.

5 The Chanterelles 99
Cantharellus cibarius, Cantharellus lateritius, Cantharellus cinnabarinus, Craterellus cornucopioides, and Craterellus foetidus

Yellow, black, orange, and red—wildly popular chanterelles are a group of vase-shaped summer mushrooms that often grow near oaks.

6 The Boletes 119
Xanthaconium separans, Strobilomyces spp., and Gyroporus castaneus

These terrestrial mushrooms have pores on the underside of the cap. There are hundreds of bolete species; here are a few easy ones to identify.

7 The Puffballs 129
Calvatia gigantea, Lycoperdon pyriforme, Lycoperdon perlatum, Calvatia cyathiformis, and Calvatia craniformis

You can stomp on them when they're old, or eat them when they're fresh.

8 Take the Field without Getting Hurt 137
Agaricus, Coprinus, Macrolepiota, and Lepiota

Wherever there's grass, look for these edible wild mushrooms.

9 Let's Eat 159
Recipes and Advice for Cooking Wild Mushrooms

From simple fried morels to a champagne breakfast with fungi, fourteen great chefs from Illinois tell you how they get wild with mushrooms.

Index 205

Preface

Is This Book for You?

This book is intended for beginners, but will also be of value to anyone interested in wild mushrooms. If you're a beginner, know that identifying wild mushrooms requires the ability to recognize certain distinctive features, such as a hollow interior or a blue cap. Although some mushrooms found in Illinois are tricky to identify, the mushrooms in this book should be easy for beginners.

For example, if you can recognize the obvious differences between a carrot and a pumpkin—that is, if their orange likeness doesn't confuse you or leave you unable to decide which is which, carrot or pumpkin—this book is for you.

About This Book

There is no way to learn forty different edible mushrooms merely by flipping through this book. It can't be done. However, let's make the best of this unexpected disappointment. If you actually read this book and pay attention, you can learn how to recognize forty different edible mushrooms found throughout Illinois and surrounding states. The details will be explained to you shortly.

But please don't skip ahead. This is important.

We understand that some of you truly know nothing, or just a little, about wild mushrooms. Still, this is your book. We created this guide for you and everyone else who wants to learn which wild mushrooms in Illinois are safe to eat, which ones are poisonous, and how to cook the good ones.

There's a real need for a field guide for amateur mushroom hunters in Illinois because there are so many people without much experience collecting and eating wild mushrooms—and that, frankly, makes us a little nervous.

There are other ways to learn how to identify wild mushrooms. One could buy several different mushroom books and spend a few years memorizing those books, consulting professionals while constantly hunting for mushrooms—and only then finally be able to confidently pick and eat fifteen or twenty different edible wild mushrooms.

But most people don't care to bother. And so we offer this little book for all of you: the shortcutters of Illinois, the skippers of instructions, the eager but impatient—the totally average, ordinary mushroom hunters of Illinois.

However, you must *read* this book—and we ask you to read it when you have the time to really pay attention, because this is serious. This is *not* a book for glancing. Identifying wild mushrooms isn't as simple as choosing one picture in a book, nodding at the accompanying words, and then heating a skillet.

We suspect you've already glanced through the pages ahead, and you've probably learned something from your preliminary browsing: this isn't a complete guide to all mushrooms known to exist. It's not even a complete guide to all Illinois mushrooms, since there are literally thousands of species out there. This book skips all of that additional information and describes here only what most people really want, which is free food.

Other Places You Can Use This Book

You've probably guessed that mushrooms found in Illinois might also occur in Iowa, Wisconsin, Indiana, Missouri, or Kentucky—states that border Illinois. It's true. For the most part, mushrooms found in Illinois also can be found in neighboring states. Not always, but almost always. If you happen to live near the Illinois border, this book doesn't become useless the moment you drive into Wisconsin or Missouri or Indiana or Iowa or Kentucky. For basic purposes, this could be a midwestern book.

But let's not stray too far. What this book gains in convenience, thanks to its focus on Illinois, it lacks in portability. This book isn't recommended for California, for example, because there is little overlap in mushroom species between Illinois and California. For instance, the poisonous *Agaricus californicus* is not known to occur in Illinois, and *Lactarius indigo* does not occur in the far western United States.

Fungi, like other living things, grow only in those habitats and climates in which they've adapted to grow. When we say that a certain mushroom found in California has never been documented in Illinois, that's in large part because the environment of Illinois does not match the environment of California. They've got a coastline bordering a saltwater ocean; we've got Lake Michigan. They've got giant redwoods; we've got sugar maples. You get the idea.

Even within Illinois, some species occur in some places but not others. Environments vary across Illinois, and biological adaptation has its limits. All of this is to say that *Edible Wild Mushrooms of Illinois and Surrounding States* is safe to read while driving across the state line, and for many miles ahead. But we recommend you put the book away after you've been driving for a very long while—or, for example, if you reach a gas station where someone stares at your license plate and says, "Illinois, huh? Well, how about that."

There are other reasons this book is best kept for use around Illinois. Not only will unfamiliar species of mushrooms appear (or familiar ones vanish), but also some species recognizable in Illinois might look somewhat different in distant regions—bigger perhaps, or smaller; brighter yellow, or maybe more brown with less white; or . . . just *different*. And with that change comes the associated risk of misidentifying a very common species you would otherwise recognize easily in Illinois.

In addition, mushroom flavors can vary from region to region. A chanterelle mushroom, for example, is not identical everywhere it grows. The chemicals that produce flavors in mushrooms can change from one environment to the next. (For more about mushroom flavors, see "About Mushroom Flavors," pp. 161–63.)

Finally, use this book with good, careful judgment. If you decide to cook and eat a wild mushroom in some far-off land, based solely on what you read in this book, know that you will be totally alone in the universe the moment you swallow whatever mushroom you actually found.

Good luck. *Read carefully*. Enjoy.

Acknowledgments

The authors gratefully acknowledge the contributions of so many who gave us everything we put into this book. Walter J. Sundberg of Southern Illinois University–Carbondale started it all, and his influence on this long-awaited and necessary project shines on every page. Thanks to the incredible chefs of Illinois. Thanks to mycologists Hal Burdsall, Patrick R. Leacock, and Roy E. Halling for their careful review of the text.

The second author thanks his favorite collecting partner, Betty Strack. Thanks to Richard J. McFarland for his assistance preparing the photographs for press. Thanks to Jay Damm, Richard Kerrigan, the Field Museum in Chicago, the Illinois Department of Natural Resources, and all the others who helped us in so many ways.

Do Not Ignore This Warning!

Mushroom poisoning can be fatal.

Deadly poisonous mushrooms are common in every region of Illinois and the surrounding states and can be found in all months—even in winter. There is no simple, one-size-fits-all rule or saying to separate the edible mushrooms from the toxic ones. It is the responsibility of the reader to pay attention and to use sober judgment while reading these pages. Neither the authors nor publisher accepts responsibility for any ill effects that might result from anyone eating any wild mushroom.

Tips for Beginners

How to Find, Identify, and Understand Wild Mushrooms

Finding wild mushrooms in Illinois is incredibly easy. They're everywhere. But that's the problem. There are so many mushrooms—and, to the untrained eye, they often all look alike, or all look different, with different colors, shapes, and sizes. Mushrooms grow in your yard, in the park, or on your neighbor's tree. They grow in wood chips beside the mailbox. Mushrooms are everywhere.

You probably have no idea how many different species of mushrooms exist in Illinois. Nobody does, really—at least, not yet. Scientists known as *mycologists* continually add to the list of known mushroom species they've documented in Illinois, and it's estimated there are hundreds of additional species—possibly species new to science altogether—waiting to be documented in Illinois.

But here's one encouraging fact. People have been picking and eating wild mushrooms for thousands of years. We're all living proof that certain wild mushrooms are, in fact, edible, regardless of those undocumented ones. This chapter covers what those millennia of experience taught mushroom hunters—how to find and identify certain well-known edible species.

1. Never Eat a Wild Mushroom You Cannot Positively Identify as Being Edible

This should be obvious—like "Never poke a fork in your eye." It's common sense. But people do stupid things sometimes. And if you think you're an expert after thumbing through this book, you're not. What this book will give you is the confidence to begin picking and eating certain wild mushrooms found in and around Illinois. But don't get cocky about it. One of the tragic mistakes people make after they've learned how to identify a few edible species is their foolish willingness to eat a mushroom they're only *pretty sure* is the right one.

Clavulina zollingeri (not edible).

Stick with what you know, positively. Look carefully at each mushroom you pick to make certain key traits match their scientific description. If not, toss it. No mushroom in the world is so delicious that people should risk their life eating a species they're only *pretty sure* won't kill them.

2. How Can You Tell if a Mushroom Is Poisonous or Edible?

There is no simple way to recognize which mushrooms are poisonous and which ones are edible. None of the basic sayings or rules of thumb you might have heard about identifying poisonous mushrooms (or edible ones) can be trusted. *None of them.*

The truth is, looks can be deceiving. Even some attractive, nice-smelling mushrooms can be deadly, and ugly mushrooms might turn out to be perfectly safe and delicious. Or vice versa. There is no one-size-fits-all rule to identify poisonous versus edible mushrooms. The only way to know which mushrooms are poisonous is to learn to recognize those species of mushrooms. Learning to recognize those mushrooms requires attention. But it can be done, and it's really no more trouble than learning to identify the edible ones in this book. It's something you should *want* to learn because it's important that anyone who collects and eats wild mushrooms also knows what poisonous mushrooms look like.

3. What Is a Mushroom?

Hygrocybe coccinea (not recommended).

You only *think* you know the answer. Most people think mushrooms are basically plants that have no chlorophyll, and mushrooms therefore feed on living or dead things. You may also know that mushrooms, unlike green plants, cannot produce their own energy from sunlight through photosynthesis. You probably remember that from science class.

But what you don't know is that mushrooms are far more than freeloading parasites or decomposers. Based on DNA and chemistry, we now understand that fungi belong in their own kingdom—the Kingdom Fungi—and are neither plant nor animal. But what's more surprising to many people is the fact that mushrooms popping out of the ground or wood are only a small part of the entire organism. The "mushroom" is the fruiting structure that produces spores, the reproductive progules that are equivalent to the seeds of plants. Thus, they are more similar to apples on the tree than to the entire tree. So the mushrooms we find on our lawn or sprouting from a log are only the tip of the fungal iceberg. Connected to the base of all mushrooms is the rest of the fungus living underground, inside wood, or on any other suitable host. That much-larger, unseen network of microscopic fungal threads is what the tree is to the apple, and it's what produced the mushrooms we see.

4. Why Do Mushrooms Exist?

Such questions actually are a subject of great importance to everybody on earth. Fungi—some of which produce fruiting bodies we call mushrooms—perform a staggering number of minor and major jobs in our environment, and they exist in many different forms. As yeasts, they produce the CO_2 that puts bubbles in our beer and give rise to bread. As producers of antibiotics, they save human lives from deadly infection. Milk could never become blue cheese without fungi, and thus Roquefort and Gorgonzola would be unknown to the world of cheese eaters. "Stonewashed" denim—that fashionably well-worn look we pay extra for in jeans—actually is the result of a fungus that has been allowed to munch on the cotton material before being washed away in factory tanks (without the help of stones, we should point out).

More important for mushroom hunters is the role of fungi outdoors. Fungi are commonly known as "nature's recyclers" because they serve as the wrecking crews that move in when (or before) plants die to dissolve and disassemble the plant tissue, thereby releasing and converting precious nutrients that can once again be used by other forms of life. Nutrients that are essential for plants to flourish aren't available in limitless supply in the soil, so plants grab what they can when they are available—and there is great competition for these elements of life. Unfortunately, nutrients extracted from the soil while a plant was alive don't magically return to the soil when the plant dies. It's why farmers have to fertilize fields artificially, inserting nutrients that were hauled away with last fall's harvest.

Well-known wood eaters such as termites or carpenter ants help disassemble a forest. But fungi play the greatest role in converting the raw material of plant structures—lignin and cellulose—into "food" for other living things. If no fungi existed, trees (which are full of nutrients) would fall over dead and lay on the forest floor for ages, along with fallen leaves, sticks, acorn caps, and every other plant cell that ever died. Marauding insects alone, even with bacteria and other microorganisms, couldn't keep up with the accumulation of plant material. Very little in the way of available nutrients would ever get returned to the soil.

Of course, that would never happen because forests would cease to grow long before that theoretical doomsday. Because all of the nutrients absorbed during the tree's life would remain locked inside the dead wood, like unopened cans of food in a pantry, new plants would starve in the empty soil. Fortunately, fungi release those nutrients as they break down old forests, and new generations grab those liberated elements of life and start the process all over again.

But fungi don't merely destroy old plant material. Many fungi often live in symbiotic cooperation with living trees, attaching themselves to tree roots, encasing the root, and/or growing among the cells of the root. The host trees welcome this amazing partnership. This mutually beneficial partnership is called *mycorrhiza* (Greek for fungus root).

Mycelium
A Visible Fungus

Mushrooms are the only portion of the fungus that we usually ever see. Mushrooms are fruiting bodies, like apples on trees—the reproductive bodies of a much larger fungus. The individual strands of fungal hyphae that grow in soil or wood (or some other host) are microscopic and too tiny to notice without magnification. But when they mass together, such as with this wood-rotting fungus, the visible haze is called a mycelium. When environmental conditions are right, mushrooms are formed growing from the mycelium—and spores are produced in these fruit bodies, enabling life to go on.

Although not all fungi do this, some species of fungi are adapted to form partnerships with the roots of specific trees. In Illinois, chanterelle mushrooms, for example, tend to be found only around oak trees. It's because *Cantharellus* (the genus of fungi that includes various chanterelle mushrooms) almost always form partnerships with the roots of oak trees—and only rarely with others. In other regions, however, different species of *Cantharellus* are adapted to live with other trees, including pines and other conifers. There might be a number of different species of fungi living with or within the roots of an Illinois tree. For reasons that aren't yet fully understood, fungi and trees link only with their chosen partners. And that partnership is one of nature's most amazing tales of botanical interdependence.

A fungal "root" called a *rhizomorph* can travel long distances, increasing the reach of a hungry fungus.

Here's how it works: Even though the sturdy roots of a healthy tree keep the tree from toppling over in a breeze, the roots aren't exceedingly efficient at supplying all of the moisture and nutrients that the tree requires to thrive. Thirsty tree roots, feeling their way through soil, encounter the tiny fungal threads of a compatible fungus, some of which connect to tree roots to greatly expand the moisture- and nutrient-absorbing abilities of the relatively fat and clumsy tree root. Roots alone can't extract sufficient nourishment to feed the giant tree above. For comparison, imagine drinking from a tiny straw: If you happen to be tremendously thirsty, it would take an eternity before you said "Ahhh."

Now, imagine drinking through fifteen or twenty straws at once—then you've really got something. That's how fungi partner with trees. Tiny hyphal threads of a fungus reach everywhere in the soil, far more efficiently than tree roots. If there's a microliter of water to be extracted between grains of soil, the fungus owns it—and so does the tree.

There's a whole lot more to this tree–fungus partnership. And trees aren't the only winners in the deal. Fungi not only give trees moisture and nutrients they've transported from nature's recycling bin, they also siphon surplus sugars from the tree roots, like humans might collect sap for making maple syrup. It's a sweet, necessary source of energy to fuel the hungry providers.

But some fungi can be quite malevolent in their quest for life's fuel, sometimes entering vulnerable tree hosts as parasites, threading their way through cells, sucking the very life from heartwood and sapwood, eventually damaging or killing the tree. This might occur even as other species of fungi are assisting the tree's roots underground.

Why Forests Disappear
How Fungi Change Everything

Some species of fungi grow only on wood and nowhere else. Why is this significant? If it weren't for fungi turning dead logs into tiny rotted particles, there would be no more forests.

Picture this: Everywhere live trees used to exist, there would be nothing but piles of dead wood just sitting there—leaves, sticks, and all—waiting for the end of the world. Slowly, insects and woodpeckers might peck and nibble away at the pile of lumber. Bacteria and other microorganisms would break down some of the structure. But new tree life would essentially halt.

Because those dead trees store the essential nutrients other plants require to thrive, a food crisis would grip the forest—unless something out there extracted those delicious nutrients from the dead wood.

Enter fungi, the organic recyclers that dissolve and process dead plant material into nutrients for other plants to utilize. Amazingly, many fungi also deliver nutrients and water to trees by linking with tree roots in a symbiotic relationship termed *mycorrhiza*. (See "Why Do Mushrooms Exists on p. 3.)

Microstoma flocossa.

Although many species of fungi live cooperatively with trees as mycorrhizas or with algae as lichens, not all interactions are mutually beneficial. Even similar species of fungi might wage chemical warfare against one another, invoking toxic battles as they compete for host territory, sometimes building impermeable walls of repellent or producing shields of chemical protection so strong and fail-safe that nothing else can survive in their presence.

Sound familiar? Many of our antibiotics today are derived from these fungal enzymes and metabolites: That dab of antibiotic cream you apply to a wound to prevent infection might be quite similar to what's produced by fungi in your refrigerator on that horrifying container of forgotten macaroni.

Or it might be similar to what's being produced by a fungus in the soil in your backyard, protecting itself so that no other life—good or bad—can join the party. The ability of fungi to produce these chemical wonders is universal—trees everywhere live in soil filled with fungi, with each species of tree conducting underground affairs with only certain species of fungi. Meanwhile, those fungi are battling or entertaining other species of forest life—such as insects and bacteria—as they conquer and barter the raw materials of life. Fungi devote much of their efforts to such interactions—and occasionally they also make mushrooms.

Xerula sp.

University of Wisconsin–LaCrosse mycologist Tom Volk got the chance of a lifetime. In 2006, Volk received a heart transplant, thanks to the organ donor program. Amazingly, fungi helped his recovery. Cyclosporine—a drug derived from a fungus called *Tolypocladium niveum*—makes organ transplants possible by suppressing the body's immunity response to new organs. Volk was home from the hospital a few weeks later, driving around LaCrosse, living it up as proof of the amazing benefits of fungi—and organ donors.

Learning which species of trees in Illinois host the species of fungi that produce edible mushrooms is a smart investment for the clever mushroom hunter. So here's a tip: Learn to identify trees. Read the habitat descriptions for each woodland species in this book, and then look for those trees where those mushrooms can be found.

5. Is This Book Really for You?

Walking outdoors and safely identifying forty species of edible wild mushrooms is possible to accomplish using this book, regardless of prior experience and without having to learn all of the thousands of species of fungi in Illinois. Many mushroom hunters recognize only one or two safe species, which is fine, as long as those people stick to what they know. This book can help you learn to identify forty different edible wild mushrooms found in Illinois, given time and study.

Just as it is possible to learn to plink out a few simple tunes on a piano without actually having to learn how to play the piano (think "Chopsticks"), ordinary people—beginners—can learn to recognize some of the wild mushrooms in Illinois without learning all of them.

Some of you might want more. The authors are quick to encourage further study of wild fungi by reading additional books, consulting mycologists, or joining the Illinois Mycological Association, which is a great organization for both amateurs and pros and is based in Chicago. Club members hold regular forays, usually with at least one expert on hand, to collect and identify all kinds of fungi. Nearby states, such as Missouri and Wisconsin, also have mushroom clubs (the St. Louis–based Missouri Mycological Society is a great option for mushroom hunters in southern Illinois). The North American Mycological Association is the parent club of most regional mushroom organizations, and membership in that group is also open to both rank amateurs and seasoned pros.

6. When and Where to Look for Mushrooms

Mushrooms grow on different things, such as on wood, in your front lawn, or maybe around certain trees, and at different times of the year. Each species of wild mushroom usually has a "season," such as spring or fall, when conditions are right for that particular mushroom to appear. It might be the only time of year that species of mushroom can be found. Early to mid-spring, for example, is the only time of year morels can be found in Illinois. Therefore, it would be absolutely fruitless searching for a morel if it happens to be June. A little late is still too late.

Because those prized morel mushrooms cannot be found in summer, fall, or winter, your morel search can be narrowed to spring only. Also, knowing that a morel mushroom usually grows around specific trees makes your morel hunting more efficient because you'll look for morels around just those trees. So, just like that, we've narrowed our morel-hunting search to one season, spring, and one habitat—near certain trees. We can apply this same information to most mushrooms.

The more questions we ask ourselves before hunting for mushrooms—the more we know about habitats and seasons—the easier it will be to look for specific mushrooms in the right place at the right time. Read the habitat descriptions for each mushroom described in this book. They'll tell you when and where to look for that species of mushroom. Of course, that doesn't mean that a search for a given mushroom in the correct season around the appropriate tree guarantees success. Mushroom hunting is like looking for treasure with only a very general map to help. The fun is in the search—as well as the find.

If nothing else, simply walking in the forest is always worth the trip, even if we fail to collect a single mushroom to eat. There are a lot of people in Illinois who would gladly collect no mushrooms at all, if only they could go for a walk in the forest.

7. Pay Attention to the Weather

Successful mushroom hunters know exactly when it rained last, how much it rained, and if it was a good, drawn-out, slow soaker or a quick, gutter-washing thunderstorm lasting seven minutes. Experienced mushroom hunters know these things because moisture is critically important in the fruiting cycle of mushrooms. One shouldn't, for example, suddenly decide to go Black Trumpet hunting without considering when it rained last. Some species of mushrooms (such as Black Trumpets) require a prolonged, soaking rain to trigger their appearance. Hunting for such mushrooms after a gentle sprinkle—or extended dry weather—usually means you'll find leaves and sticks, and maybe some chiggers, but no mushrooms, even if you are looking in the "right" spot.

8. Is It Harmful to Pull Them Out of the Ground?

Of course you want mushrooms to grow back in the same place where you just picked a basketful of real beauties. A mind inclined toward conservation should not be discouraged. But, to be honest, it probably doesn't matter if you pull them out of the ground or gently slice them at the base. Remember, mushrooms are the fruiting bodies of a much larger, longer-lived fungus that's attached to the mushroom you're about to pick. And fruiting bodies—such as apples on trees—have evolved to survive being harvested by hungry passers-by.

When, for example, we pull an apple from a tree, sometimes we pull a twig or two from the branch as we jerk the apple into our possession. You've probably done it. Notice how the tree always survives without a problem? Fungi are like that. The slight bit of fungus you might disturb while removing its fruiting bodies is like the leaf on your apple stem. It's no big deal.

All of this cutting-versus-pulling anxiety stems from the notion that mushrooms won't return next year if we damage the delicate little "roots" beneath them. Basically, forget all of that. Being mauled by a hungry mycophagist is nothing new to fungi. Squirrels regularly pull or dig mushrooms out of the ground. Truffles wouldn't persist if something didn't dig them up and distribute their spores. But the reason people should mostly avoid yanking mushrooms out of the ground has nothing to do with harming the fungus; it has to do with clean food. Too many people collect mushrooms by pulling them out of the ground and then tossing the mushroom and a clod of dirt into a plastic bag. It's like pulling flowers out of the ground when all we really desire is what's above the soil. Back in the kitchen, you will spend far less time cleaning mushrooms if the mushrooms arrived in the kitchen fairly clean.

Of course, certain mushrooms *must* be excavated to be certain of their identification. Deadly *Amanita* mushrooms rise from an egglike sac just beneath the soil surface (see the photo on p. 11). When picking any mushroom that vaguely resembles an *Amanita* (very young, unopened caps of Meadow Mushrooms, for instance, merit the digging treatment), it's helpful to scoop the stem base from the soil to be certain the remnants of that sac—which would indicate an *Amanita*—are not hidden underground. Even as we do such precautionary digging, the fungus that created that mushroom happily carries on.

So pull them out of the ground if you honestly don't care at all about dirty mushrooms, or slice them carefully, all tidy and neat, if you're a clean freak. It won't affect the fungus or the chance that you'll find the fungus again next year.

9. You Will Get Very, Very Discouraged

A supportive book would have advised something different, such as "Don't get discouraged, kid" or "Please be patient."

But we're realistic. The fact is, you will get maddeningly discouraged sometimes while hunting for wild mushrooms. Sometimes, mushrooms just aren't growing where you expect to find them. And it's not your fault. Even the best mycologists—scientists who study mushrooms—return from the woods once in a while and claim they were merely looking at trees.

Mushrooms appear when they appear, and they do not appear at any other time. And while the occurrence of many mushrooms can be quite predictable, like a holiday, mushrooms are also capable of not doing what we think they're going to do.

You'll get used to it. But, with time, you'll also get better at predicting when and where to find a certain mushroom. After several years of collecting wild mushrooms, you will gain enough experience to walk outdoors with the confidence of a braggart—pointing your bat toward the bleachers, so to speak—examining your nails in boredom, knowing without a trace of doubt you will find a tasty species fruiting exactly where you're headed, knowing it will be perfectly fresh that day because you've found it in that very spot every other year at exactly that time.

The egglike sac of the Destroying Angel (*Amanita bisporigera*).

Occasionally you will hit the nail on the head—enough times, perhaps, to gain you the enviable reputation of being the local authority on mushrooms. But often you will be frustrated. The sought-after species you've found so many times in habitat exactly like where you're looking right now simply won't be there, and you will be embarrassed. People will abandon you, and say you're not an expert after all.

Maybe, eventually, that mushroom you sought will, indeed, appear sometime later, and in that exact spot. It might happen next year, or maybe not—it's all part of the experience of hunting for wild mushrooms. With experience, you'll accept the fact nature is notoriously fickle when it comes to predictability, and mushrooms exemplify that unpredictability.

Unlucky hunters have an expression. They say, "That's why they call it hunting, not taking." And that's how it really is.

10. Examine Any Kind of Mushroom You Find

There is so much to learn from all mushrooms—why examine only the edible ones? Just because a mushroom has a lousy flavor or is inedible doesn't mean it's a worthless mushroom. When we pause to inspect whatever mushroom happens to be in our path, we begin to understand the possible variations of shape, texture, and "behavior" of all mushrooms.

Yes, this book is all about edible mushrooms. But when we tell you a particular edible mushroom has gills that are not attached to the stem, it's helpful to discover a mushroom—edible or otherwise—that actually has gills growing out of the side of the stem beneath the cap to better understand the description being made.

Many mushroom hunters walk straight to their patch of morels without hesitation, anxious to review the crop. Patience would seem a superfluous virtue. But there is much to be learned about the secret life of your morels if you pay attention to other fungi you might encounter along the way.

Yellow Chanterelles.

11. Keep Notes

Chances are, you don't remember exactly what those still-tiny leaves on pawpaw trees looked like that day last year when you found morels growing in the woods over by the mall. You could've jotted it down, you know. People have been keeping secret, personal diaries of their mushroom discoveries for years. It helps them remember all of the juicy details of when and where their favorite mushrooms appeared in their neighborhood, not just in some book.

The Secret Life of E. C. Branson

Everybody has a secret life: Politicians. Celebrities. You.

U.S. history recalls Eugene Cunningham Branson (1861–1933) as a college president and pioneer in rural economics and sociology. He traveled to Europe to study rural economics; he wrote a book about it and even advised U.S. presidents about rural economics. He became a famous humanitarian—Booker T. Washington knew his name. E. C. Branson devoted his life to improving inefficient rural economies. It was his claim to fame.

Still, everyone has side interests, and Branson the famous humanitarian wasn't opposed to studying his neighbors with a different eye. Some time after 1923, Branson acquired a copy of *The Mushroom Book* by Nina L. Marshall and began keeping notes in the page margins.

All mushroom hunters can improve their fortunes by keeping good notes whenever they find what it is they want.

On September 16, 1928, the "Coker place" had what E. C. Branson wanted.

E. C. Branson.

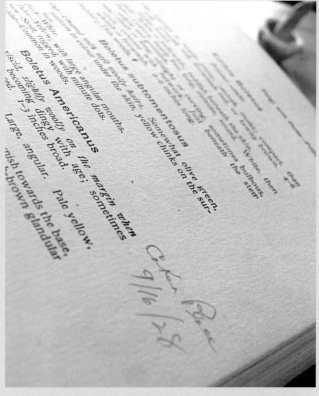

E. C. Branson's 1923 edition of Marshall's *The Mushroom Book*, with observation notes.

Grifola frondosa.

Laetiporus sulphureus.

12. Why Do Mushrooms Look Different?

Why do fish look different? Or birds? Everything looks different, really. Every species does different things in this world; they live in different places and eat different foods. They're adapted to where they live.

Why a mushroom develops orange brackets instead of swelling out like a white balloon is just one of those things mushroom hunters don't have to know. What mushroom hunters should know is that those features help identify mushrooms, and recognizing those features really can be a matter of life or death.

Coprinus comatus.

Lycoperdon perlatum.

Hericium americanum.

13. What Diagnostic Traits Identify Mushrooms?

Most mushrooms seem to look alike until we learn what characteristics to look for. Surprisingly, the most important characteristics to note when trying to identify a mushroom aren't always the most obvious features we instantly see, such as the overall color of the mushroom or its shape or size (although those traits are, indeed, important). Does the mushroom have gills under the cap, or does it have tiny pores instead of gills? Does the mushroom lack a proper cap due to a completely different growth form, such as a bracket on a tree or a round puffball? Many mushrooms are just too weird-looking to compare them to anything else at first glance. Therefore, one must play fungal detective and ask key questions that eventually rule out all but the correct species.

Where and when a mushroom was found can be equally important because many species of edible mushrooms grow only in certain habitats and only at certain times of the year. Was it growing on wood or on the ground? Was it found in April or August? What color are the spores—green, brown, white, or pinkish? (See "How to Make a Spore Print," p. 16.) Does it have a ring on the stem? Is there any evidence beneath the stem that the mushroom popped out of an egglike sac underground? Does it change colors when bruised or exude a liquid when cut—and, if so, what color is the liquid? Does it have a noticeable smell, perhaps like unripe watermelon rind or bread dough?

These are some—but not all—of the basic features you'll want to observe when trying to identify mushrooms. Each edible mushroom described in this book has an accompanying list of features you should compare with whatever mushroom you found. If the traits don't match perfectly, don't eat it (not even a little bit, as a test). Don't be foolish.

14. Is It Legal to Collect Mushrooms Everywhere in Illinois?

Edible mushrooms grow everywhere, but they don't always grow on your property. You might spot a huge Chicken Mushroom, for example, perfectly fresh and ripe for picking—but it's growing on a tree in the yard of a stranger. Oyster Mushrooms might tempt us with vast bunches clustered on a willow in the city park, totally forbidden. Mowed lawns around fire departments or in front of City Hall might sprout a sea of *Agaricus*, and nobody will be watching. Of course, they're not *your* mushrooms, technically.

We would like to list every Illinois city ordinance and park district rule that applies to the collection of wild fungi, but we can't. There isn't space. Start with the fact that private property is private property. Also, public property is not always "public" property. More often than not, it's illegal to collect mushrooms in your local forest preserve (the Cook County Forest Preserve, for example, requires a special collection permit, which is usually reserved for scientific purposes), but collecting edible wild mushrooms is perfectly legal in any Illinois state park (but not in designated nature preserves within those state parks). Although collecting edible fungi in Illinois state parks is legal, mushrooms collected there cannot be offered for sale.

State, federal, and local laws regarding the collection of fungi aren't identical, meaning it might be legal to collect in one public forest but not another. Regulations do change over time. Perhaps with raised public awareness of the renewable resource of edible wild fungi in Illinois, site regulations can be modified to allow all residents the pleasure of collecting edible wild mushrooms at their own risk on public lands that are not otherwise restricted. But we are not there yet.

What about Those Other Mushrooms?
Unfortunately, You Can't Eat Them All

The late astronomer Carl Sagan used to speak enthusiastically about the billions and billions of stars in distant galaxies, as if any one of them might hold the amazing secrets to the origin of life or the answer to every question ever wondered. Every star was extremely special to Sagan, yet he also realized almost none of those specks in the sky held the faintest potential of being inhabited by even basic life.

Still, they were all amazing, every single speck.

One might consider the well over one thousand species of mushrooms in Illinois like Sagan considered stars—each one is absolutely amazing. Some mushrooms bioluminesce; some knock out flies; some dissolve in a matter of hours. Mushrooms really are incredibly cool, even if only a tiny fraction of them are good to eat.

Lots of mushrooms are difficult to identify—including many common mushrooms. This mushroom belongs to the genus *Russula*. Hundreds of different *Russula* species exist in North America. Some are toxic.

How to Make a Spore Print

Making a print from the dustlike spores dropped from a mushroom cap is one of the important methods used to help identify various mushroom species. Some mushrooms have white spores, or rusty brown, green, or black spores—or nearly any other shade of almost any color.

Sometimes the color of the spores separates otherwise hard-to-tell-apart species. The edible Meadow Mushroom, for example, can be separated from a deadly *Amanita* by the fact that Meadow Mushrooms have chocolate brown spores, and all *Amanitas* have white spores. Making a spore print really is easy, and it can help reassure beginners that they've made no mistakes in identification. Every wild mushroom hunter should learn how to make a spore print. The process is simple.

Remove the cap and place it over white and dark paper (to show light or dark spores). Cover with a bowl, cup, or other object to keep wind off and humidity high. Wait at least six hours, and then lift up the mushroom cap to study the color of the spores it dropped.

The olive green spores of the poisonous Green-Spored Lepiota.

15. Are Mushrooms Good for You?

Food is good for you. It keeps you alive. If you didn't eat, you'd be history. But just how important are mushrooms among that pile of food? Is the nutrition we get from a mushroom unique among the ingredients we eat daily to remain alive? Do mushrooms offer us any *special* ingredients to make us live even healthier lives?

Honestly, you could live without mushrooms. But you could also live without chocolate, or steak, or milk, or any other food item whose primary nutrients can also be obtained through some different food source.

The question of whether individual species of mushrooms might contain special nutrients or healthful boosters really should first acknowledge the reality that all food contains magic ingredients. Limes, for example, once saved sailors from scurvy. Thanks to the magic ingredient Vitamin C, which really isn't a magic ingredient at all—no more than protein is a magic ingredient in meat—sailors forgot all about scurvy. Both protein and Vitamin C—and a slew of other ingredients—mean the difference between life and death for our omnivorous human metabolism.

But do mushrooms offer any *really* special health improvers? Or cures? Unfortunately, the answer here also redefines the question. When we inspect the subject closely, really considering what all of that eaten food actually does inside of our bodies, we realize food is basically a swarm of business transactions waiting to happen in our stomach. There are sugars and starches and enzymes and pathogens and toxins. All of them potential fuel or foe, ready for action.

"Food" really is an enigmatic collection of natural resources. Mushrooms, like everything else on your plate, offer a staggering number of intentions and possibilities, with each human body responding differently to all of that potential. The substance *coprine,* for example, might visit your bloodstream without noticeable effect after a meal of mushrooms and you'll live happily ever after. But if you happen to drink alcohol within a few hours or days after eating food containing coprine, you might suddenly feel nauseously ill, or worse. That's because coprine—an amino acid found in just a few mushrooms—prevents the breakdown of alcohol. If coprine is present in the bloodstream when alcohol rolls in, the party always ends in the bathroom.

Then there's a galactomannan that is found in Yellow Morel mushrooms (*Morchella esculenta*). The galactomannan is a rather complex bunch of sugar molecules known as a polysaccharide, and Yellow Morels produce a pretty good amount of it, roughly 2 percent of their dried weight. That means a typical serving of several Yellow Morels will give you perhaps twenty milligrams of galactomannan.

Big deal? Well, the polysaccharide in Yellow Morels happens to be an active immune system stimulant. It can trigger an immune response in humans, which can be a good thing. A polysaccharide isn't a magic ingredient, no more than Vitamin C is a magic ingredient, but it just happens to be present in Yellow Morels.

So, back to the original question: Are mushrooms good for you? Well, they can be. But since food can be good for you, we're pretty much back where we started.

2

Common and Poisonous

A Few Toxic Mushrooms

Some species of poisonous mushrooms appear so frequently around Illinois that we think they deserve special mention. If your goal is to collect only edible mushrooms, it helps to know what the most common poisonous mushrooms look like.

The next few pages describe some of the historically common poisonous mushrooms found in the Prairie State, as well as the potentially poisonous *Amanita thiersii*, once considered an obscure southern species that began turning up everywhere in southern Illinois in the late 1990s and quickly became one of the most common "bad" mushrooms found on Illinois lawns. This isn't a complete list of toxic mushrooms in Illinois, just the most commonly encountered ones. There are thousands of different species of mushrooms out there in the world, and while a number of them are good to eat, many others have never been tested for toxicity.

Learning to recognize a few different poisonous mushrooms will help you become a more skillful mushroom hunter, just like learning to recognize a rattlesnake or a black widow spider will help you survive outdoors. Knowing one's enemies—at least the common enemies—is handy knowledge. If you're truly interested in collecting wild mushrooms to eat, the fact that there are deadly poisonous mushrooms everywhere in Illinois should concern you enough to make you want to learn how to distinguish them.

What Is Mushroom Poisoning?

"Mushroom poisoning" and its effects on the human body are more complex than most people realize. Poisoning from mushrooms is not a simple, universally identical medical condition, no more than "being sick" always indicates someone has contracted West Nile virus.

There are different kinds of mushroom toxins known to cause bad reactions when ingested. Some mushrooms contain toxins that attack the liver and kidneys, while some are severe gastrointestinal irritants. Sometimes, normally edible species growing in toxic places might contain dangerous chemicals or elements extracted from those contaminated soils or other toxic substrates (poisonous substances that might quietly accumulate in the body until an "overdose" triggers symptoms of poisoning, which can be fatal). Some toxins, such as coprine, are generally harmless for most people unless alcohol is consumed within a few days of eating the mushroom. Other toxins, such as ibotenic acid, affect the nervous system and can produce anxiety, nausea, hallucinations, and convulsions.

To complicate matters, some people become ill after eating mushrooms other people might eat safely. Mushrooms can be like peanuts or shellfish or milk or any other kind of food item some people—but not all people—can eat without adverse reactions. Two people might share a meal of the same species of mildly poisonous mushroom, eating similar quantities, and one person will end up in an emergency room, having the contents of their stomach evacuated, while the other person might be fine, staying home to finish uneaten leftovers.

Deadly poisonous mushrooms are another matter. Nobody can eat deadly poisonous mushrooms without suffering severe—often deadly—consequences. Do not experiment with even small amounts.

Green-Spored Lepiota
Chlorophyllum molybdites

No other mushroom in America causes more poisonings than the Green-Spored Lepiota. It's because this warm-weather lawn resident tempts reckless mushroom lovers with its large, scaly white caps and reportedly pleasant flavor. It looks edible. Some people mistake it for the edible Parasol Mushroom (*Macrolepiota procera*, p. 150), but that's because they missed an obvious clue. The gills of this mushroom turn dingy gray green with age due to the color of the spores, which are a dirty olive green. So you'll want to lay a Green-Spored Lepiota cap on a piece of paper to make a spore print (see p. 16). The resulting spore "dust" will reveal the color of the spores, even if the young gills still appear white.

Never eat this mushroom.

There is a ring on the stem, which might slide freely or break off and fall to the grass. When young, the mushroom cap is round and resembles a golf ball on a stick. You will probably see this mushroom a lot every summer for the rest of your life. It can be quite common and very easy to identify after making a spore print. It really is an easy-to-recognize enemy.

Green-Spored Lepiota (poisonous).

COMMON AND POISONOUS

GREEN-SPORED LEPIOTA FEATURES

CAP: 3–10 inches across when fully expanded; it's one of the largest caps of any lawn mushroom found in Illinois. (Every summer, many people report mushrooms "the size of dinner plates" on lawns.) White and globular when young—like a golf ball on a stick—usually with a tan to brownish center, often peeling in patches. Loose, scaly, tan (not reddish) flecks scattered on cap.

GILLS: White when young, not attached to the stem, turning drab olive gray with age as the olive green spores mature. The gills can remain white for a long time, making the color of the gills an unreliable identification feature.

SPORE PRINT COLOR: Olive green (see "How to Make a Spore Print," p. 16).

STEM: Sturdy, smooth, without brown speckles or scales. A light sheen is often present. Color can be variable shades of silvery gray to greenish to purple gray. When bruised, a yellowish green cast slowly appears. A loose ring, which sometimes falls off, should be prominent.

HABITAT: In grassy areas during warm weather, often growing in rings or arcs. Common.

COMMENTS: Why do so many people get poisoned by this well-known poisonous mushroom? Several reasons. The Green-Spored Lepiota is a common summer find on front lawns where kids, pets, and unwitting adults decide to sample the attractive—and reportedly tasty—mushroom. It has been mistaken for the Parasol Mushroom (p. 150) as well as a related species not recommended for beginners called the Shaggy Parasol (*Macrolepiota rachodes*). Poisoning symptoms commonly include severe intestinal discomfort, vomiting, diarrhea, with fairly rapid onset—usually within a couple of hours after eating this mushroom. Some people reportedly eat it without ill effects. Although the Green-Spored Lepiota is a toxic mushroom, it isn't known to cause death.

Scattered colonies of Green-Spored Lepiotas, often in rings or arcs, appear on lawns in warm weather.

The strong, fibrous stem, sometimes curved, often shows a reflective sheen. Compare against the brown-speckled stem of the Parasol Mushroom (p. 150). When young and freshly opened, the Green-Spored Lepiota has white gills. As the mushroom matures, the green spores are released, dusting the gills sickly gray green. That color helps separate it from all other Lepiotas—some of which are edible. But this Lepiota—the Green-Spored Lepiota—is poisonous. A spore print (p. 16) will help identify this stomach-wrenching terror found on front lawns everywhere.

With age, the gills of the Green-Spored Lepiota turn dirty olive green.

In wet weather, the features of the Green-Spored Lepiota can be distorted or exaggerated by swelling.

COMMON AND POISONOUS

Deadly Galerina
Galerina autumnalis Group

At first glance, who would ever guess these small, brown-capped mushrooms growing on logs everywhere can be deadly poisonous? You've probably seen them. They're especially common in cool weather. And they're totally ordinary looking. But the Deadly Galerina happens to contain amatoxins—the same lethal chemicals produced by deadly *Amanita* mushrooms.

The poisonous Deadly Galerina often grows on rotting logs.

Yes, appearances can be very deceiving.

Due to the risk of potentially fatal poisoning, you'll want to pay very close attention when collecting any small, gilled mushroom found on wood. Never confuse the edible Velvet Foot (p. 53) with this mushroom. Also compare with Oyster Mushrooms (p. 46) and Honey Mushrooms (p. 67).

DEADLY GALERINA FEATURES

CAP: Small, often no larger than 1 inch across. Rusty brown to reddish brown, rather slimy looking and tacky when wet. Edge of cap curls over gills when young (a membrane covers the gills at this young stage). Basically smooth, dome-shaped, often with traces of the silvery membrane lingering on the edge of the fully opened cap.

GILLS: Light to dark brown, attached to the stem but not running downward onto the stem. Basically the same color as the cap, perhaps a bit more honey colored.

SPORE PRINT COLOR: Rusty brown (see "How to Make a Spore Print," p. 16).

STEM: Silvery gray to brown, roughly scaly when young. A membrane that covers the gills when young remains attached to the stem and forms a thin ring or collar after the cap expands and breaks the membrane. The ring is sometimes indistinct, but because these mushrooms often grow in large troops, some of the fruit bodies will likely have a ring.

HABITAT: Most common on rotting logs, stumps, and wood mulch, especially wood in advanced stages of decay. Very common in cool weather, including the dead of winter.

COMMENTS: There are several species of very toxic *Galerina* mushrooms known worldwide. They all basically look like this one, and no further description of those very similar species is needed as long as mushroom hunters avoid anything resembling a *Galerina*. The edible Velvet Foot (p. 53), which also grows on wood during cool weather, differs from the Deadly Galerina in a few ways: The Velvet Foot doesn't have a ring on the stem, its spores are white, and the fuzzy stem that darkens toward the base are traits that collectively separate it from the Deadly Galerina. Honey Mushrooms (p. 67) can be differentiated by their white spores and fibrous stem. Know that different species of mushrooms can grow side by side, like weeds in a vegetable garden, and it's entirely possible a poisonous *Galerina* could poke out among a cluster of edible Velvet Foot or Honey Mushrooms. As a general rule, be very wary of "LBMs" (little brown mushrooms).

The ring on the stem of Deadly Galerinas can be easy to notice—or very faint.

The Deadly Galerina has rusty brown gills and spores.

COMMON AND POISONOUS

Destroying Angel
Amanita bisporigera

The snow white appearance of the Destroying Angel is a good example of nature's deceptive practices. Inexperienced mushroom hunters might guess this beautiful, pure white mushroom, with its graceful skirt around the stem, is totally harmless. But exactly the opposite is true. One cap from *Amanita bisporigera*—or its twin species, *Amanita virosa*—might contain enough amatoxin to kill a healthy adult.

Like all species of *Amanita,* the Destroying Angel begins life encased in an egg-like membrane, just below the soil surface. As the mushroom pops out to grow, remnants of that cuplike membrane (known as a *volva*) persist at the base of the stem, but mostly underground. Dig carefully beneath the stem to find that evidence. No edible mushroom included in this book should have a volva at the base of the stem. Also, never confuse the partially buried "egg" of an immature *Amanita* with a small, edible puffball (see "The Puffballs," p. 129, for comparison).

DESTROYING ANGEL FEATURES

CAP: Medium-size, 2–3 inches across when fully expanded; white, satiny smooth, usually free from flecks or warts. When young, the entire thimble-shaped cap is encased in an egglike membrane called a *universal veil*. As the cap matures and expands to full size, it flattens into a white disk.

GILLS: White, fading with age to dirty white—but still mostly white. Gills are not attached to the stem (i.e., they are free). When young, the gills are covered by a thin, white membrane that tears away as the cap expands, leaving a collar or skirt on the stem.

SPORE PRINT COLOR: White (see "How to Make a Spore Print," p. 16).

STEM: White, smooth, tapering upward toward the cap. A thin ring or collar (or its remnants) should be visible. The swollen base arises from a sac just below the soil surface, although the delicate sac might deteriorate with age—careful digging should reveal this feature.

HABITAT: Grows from soil around trees, usually oaks, from early to late summer. Often found solitary, or in a small group. Found throughout all of the Midwest and much of North America.

COMMENTS: All parts of this mushroom should be white or faded slightly with age. The critical diagnostic traits include the collar or skirt around the upper stem (a thin membrane that might deteriorate with age), the egglike sac at the base of the stem below the soil surface, gills that are not attached to the stem, and a white spore print. This is one of the deadliest mushrooms known to exist. No simple antidote exists and death is possible unless immediate medical attention is sought after eating as little as one cap. No method of cooking or boiling can destroy amatoxins. Toxicity might vary from one mushroom to the next, and while some people manage to survive amatoxin poisoning through dialysis or even liver and kidney transplants, such drastic measures are certainly not an easy cure.

A white membrane covers the white gills of the Destroying Angel when young, but tears away from the cap and remains on the stem as a skirt or ring in maturity.

All *Amanitas* begin life in a saclike "egg" just below the soil surface.

COMMON AND POISONOUS

Jack O'Lantern
Omphalotus illudens

Although poisoning from the Jack O'Lantern mushroom typically results in severe gastrointestinal agony, intense vomiting, chills, and sweating, this mushroom isn't considered deadly poisonous. It is, however, a strongly toxic mushroom and should be collected for one reason only: The gills of the Jack O'Lantern glow, making it a very fun fungus indeed.

For biological purposes not yet understood, the gills of *Omphalotus illudens* display what's known as *bioluminescence* (think of fireflies). You can see the show when fresh specimens are brought into total darkness. Illinois specimens tend to be rather weak in their display, but allowing one's eyes to become totally adjusted to absolute darkness can improve one's ability to see the eerie blue-green glow from the gills.

People have mistaken this mushroom for the Yellow Chanterelle (p. 100) or even the Chicken Mushroom (p. 138). No such mistakes should ever occur if one learns to recognize the obvious features that distinguish those mushrooms.

The razor-thin gills of the Jack O'Lantern.

JACK O'LANTERN FEATURES

CAP: Pumpkin orange, sometimes with brown streaks or patches; basically smooth, without scales or warts, growing as large as 9 inches across, but often 2–3 inches. The edge of the cap is typically wavy and the center is usually depressed, especially in maturity.

GILLS: Orange, thin, and well-defined. The tendency of the gills to run downward onto the stem is why some people confuse this with the Yellow Chanterelle (p. 100), but the Yellow Chanterelle has blunt gill ridges, nearly like corduroy, as opposed to the razor-thin gills on a Jack O'Lantern. The gills are the only part of the Jack O'Lantern that glows.

SPORE PRINT COLOR: White (see "How to Make a Spore Print," p. 16).

STEM: Long in comparison to cap size, tapering toward the base, often in clusters arising from a central base. The strong stems of the Jack O'Lantern are fibrous and wavy, sometimes thicker or swollen in the middle. When fresh, stems often exude a clear orange liquid some people find irritating to skin.

HABITAT: A root parasite that grows from soil around trees and stumps, usually oaks, mainly in autumn but also in spring or summer. Rarely found solitary, usually in clusters.

COMMENTS: Inexperienced mushroom hunters continue to mistake this poisonous mushroom for the Yellow Chanterelle. The color of the Jack O'Lantern is quite attractive, and the graceful curves of the cap suggest similarities with the curves of Yellow Chanterelle caps. But the thin gills beneath the cap of the Jack O'Lantern look nothing like the blunt ridges under the cap of a Yellow Chanterelle. The stems of the Jack O'Lantern show another difference: A typical cluster of Jack O'Lanterns can be pulled from the ground as a group, linked together with a rootlike base. Yellow Chanterelles might be found scattered in colonies with many of them closely associated, but they never grow in dense clusters or bouquets of a dozen or more crowded-together mushrooms. Although Smooth Chanterelles (*Cantharellus lateritius*) (p. 104) sometimes appear in colonies with stem bases lumped together, the generally smooth, gill-free undersurface of that species looks nothing like the well-defined gills of the Jack O'Lantern. Also, the tapering, pointed base of the Jack O'Lantern stem is a diagnostic trait. The base of Yellow Chanterelles never tapers to a rootlike point.

Young specimens of the Jack O'Lantern pose the greatest risk for misidentification or confusion with Yellow Chanterelles (p. 100) because of the curled-over cap and the gills that run downward onto the stem. Yellow Chanterelles do not share the crowded, clustered growth typical of this poisonous species and do not have razor-thin gills.

The Jack O'Lantern grows above roots, beside trees and stumps, and often forms dense clusters with caps as large as 9 inches across. The cap is pumpkin orange but might also show darker patches or blotches.

Thiers Amanita
Amanita thiersii

This unusual *Amanita* grows almost exclusively in grassy areas and might pose a risk for poisoning pets, children, and unsuspecting adults because of its habitat. Most *Amanita* species grow around trees and therefore are common in forests—not in the middle of a front lawn. This mushroom created something of a sensation among Illinois mycologists when it suddenly began appearing everywhere in southern Illinois during the late 1990s. Prior to that, it was considered an obscure southern species. As of 2008, it was known to occur as far north as central Illinois, and might quickly become common statewide. This may be an example of a species moving northward in response to climate change.

The shaggy, cottony fluff coating the cap and stem make this an easy-to-identify mushroom, although that diagnostic evidence might wash away in the rain, leaving a smooth white cap and stem. The gills are white and not attached to the stem. For comparison, the edible Meadow Mushroom (p. 139) has pink gills that turn chocolate brown with age. Although this species does not produce deadly amatoxins, it is very poisonous.

THIERS AMANITA FEATURES

CAP: White, becoming dingy yellowish in age, 3–5 inches across, covered with a thick, powdery mass of loose, cottony fluff that washes away easily in the rain. When young, the white cap is round and resembles a golf ball on a stick (a trait shared with the poisonous Green-Spored Lepiota, p. 21). With age, the cap opens to a flat disk.

GILLS: White, not attached to the stem, covered at first by a cottony membrane which breaks away as the cap opens and matures.

SPORE PRINT COLOR: White (see "How to Make a Spore Print," p. 16).

With age, the cottony fluff of *Amanita thiersii* hangs in ragged scraps—unless rain has washed it off.

STEM: White, sticklike, and covered with masses of cottony white fluff that might disappear in the rain. The base of the stem rises from a partially buried "egg" membrane that covers the entire mushroom before it rises from the soil. Digging at the base of the stem can reveal evidence of what remains of this membrane.

HABITAT: Grows from soil in grassy areas during mid- to late summer in southern and central Illinois, continuing into early fall. Locally common and often massively abundant.

COMMENTS: As of the time of printing, it was not yet known what toxins this mushroom possesses. But concerned parents and pet owners should understand that deadly poisonous mushrooms aren't poisonous to touch. Mycologists collect and handle this mushroom often, examining its features, without ill effect. Although licking one's fingers after handling poisonous mushrooms isn't recommended, the notion that a child could die after picking one (or breathing its spores) is a fabulous exaggeration of actual risks. One can safely pick this mushroom to learn about its traits, and then dispose of the mushroom, without ill effects.

Felicia Bart, 94, Missouri mushroom hunter.

3

Into the Forest

Mushrooms Found with Trees

Mushroom hunters spend more time in the woods than anywhere else. That's because trees and mushrooms go together in so many ways—and mushroom hunters know it. Mushrooms grow on dead wood. They grow on living trees. They grow above tree roots. What mushroom hunters probably don't know is this: Without fungi, most trees would die for lack of water and nutrients.

About 130 years ago, scientists figured out that microscopic threads of fungi latched onto certain tree roots and formed life partnerships essential for both plant and fungus. This underground spiderweb of fungal threads is what delivers key nutrients and water the trees can't easily extract or manufacture alone. The fungi, in exchange, having no internal energy source, siphon some of the tree's plentiful sugars as food. This relationship is called *mycorrhiza*. If mycorrhiza seems to be a disingenuous partnership, with both friends picking the other's pocket, compare mycorrhiza with the relationship humans have with trees and mushrooms.

Not all forest fungi play the mycorrhizal game. Some fungi conduct other business with trees, living and dead. Consider this: When a tree falls or its leaves drop, or when annual plants wither and die, their cellulose and lignin corpses contain nutrients those plants used while living, but obviously no longer need. Until something comes along to tear apart the dead tissue and extract those nutrients, the wealth of food remains locked up like so many unopened cans in a pantry.

Fungi play that critical recycling role, breaking down plant material with the aid of specific enzymes, dissolving dead cells and extracting precious elements for the benefit of themselves and new forest life. It's a complex series of steps that reduces a dead forest to basic molecules. No one single fungus does it all; in fact, individual species of fungi show up to do their work only at certain places and stages of decomposition.

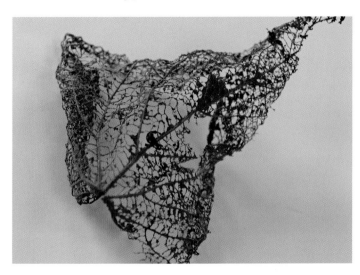

One fungus might work on the stems of dead walnut leaves, and another might dissolve the thin tissue between the "veins" of dead leaves. A forest might host thousands of species of fungi, each performing a job others do not. Some fungi compete against other fungi with brutal chemical warfare for the same resources, while others ignore the meaningless presence of a fungal relative.

Some of these wood-loving fungi produce edible mushrooms. Knowing which of these fungi associate with what trees, living or dead, is enormously helpful when walking into the woods to find edible mushrooms. This chapter includes more than two dozen edible mushrooms found in association with wood. Pay close attention to the descriptions about their habitat, and then learn to recognize a few basic trees. With a little experience, you can become one of those great mushroom hunters who steps into a forest and almost instinctively knows where to begin looking for edible mushrooms. People will admire your amazing knowledge—and knowledge is real power—delicious power.

Hen-of-the-Woods
Grifola frondosa

September—that's the traditional month to start looking for Hens-of-the-Woods. Almost like clockwork everywhere in Illinois, usually by the first frost, one, two, or three clusters of these tender polypores appear at the base of chosen oaks, growing above buried roots. They can be huge, but the gray to dusty brown caps blend in surprisingly well amid early autumn colors, which is why people have to hunt for these mushrooms. You might find five in one day. Or one every five years.

Be prepared to open your arms. This disorganized maze of overlapping lobes can be monstrous—it's one of the largest edible mushrooms you will find in Illinois. When it's really good and fresh, collect the whole thing. Everything but the thicker interior chunks are tender enough to be edible.

HEN-OF-THE-WOODS FEATURES

CAP: Overall, the mushroom includes many "caps" of various shapes and sizes in the form of interconnected lobes. Collectively, the entire mushroom might be as big as a basketball, or bigger. Upper surface of each lobe is smooth to slightly irregular; mottled or streaked with variable shades of brownish to smoky gray. Undersurface is bone white to cream—never staining dark when bruised—and is covered with very tiny pores. Individual caps can be compared to the shape of a ginkgo leaf.

FLESH: Meaty and firm but tender when fresh; whitish cream to yellowish cream. Odor mild, mushroomy, but not distinctive.

STEM: There is no central stem to speak of; the entire cluster of caps is attached to the soil by a dense base, vaguely like a head of cabbage that seems to rest on the soil. From that starting point, the mushroom grows up and outward in a confusing, interconnecting mass of lobes.

Look for Hen-of-the-Woods to grow in clumps above the roots of oaks during autumn. You might find more than one cluster beside one tree. Four Hen-of-the-Woods appeared beside this oak in Will County, Illinois.

HABITAT: Grows from the soil at the base of living or dead oaks and stumps, particularly black or red oaks, in early to mid-autumn. Usually located directly beside the tree or only a few feet away.

COMMENTS: The overlapping, labyrinthlike cluster of caps, which individually resemble the fanlike shape of ginkgo leaves, is surprisingly well camouflaged amid the distracting fall colors. In spite of the fact that a typical Hen-of-the-Woods might be larger than a basketball and weigh as much as 10 pounds or more, hunters of this inconspicuous mushroom always walk slowly through the forest. Because of the indeterminate nature of its growth, whatever object happens to be in its way as the outer margins grow gets "swallowed" or surrounded by this peculiar fungus. Sticks, leaves, and blades of grass might turn up embedded and incorporated into the flesh, hopelessly immobilized—the equivalent of boots in a bucket of concrete. It makes cleaning this mushroom trickier; seemingly perfect specimens might be sliced into pieces only to discover a twig—or a leaf or two of poison ivy—locked inside. Black beetles love this mushroom as much as we do, so don't hesitate to dunk this funhouse of hiding places in lightly salted water to evict the unwelcome.

The mass of overlapping lobes of a Hen-of-the-Woods can resemble a pile of autumn leaves at the base of an oak.

Although the pore-covered undersurface of Hen-of-the-Woods is always white, the upper surface can be variable in color: mostly gray or mostly brown—or a combination of both colors.

HEN-OF-THE-WOODS LOOK-ALIKES

The Black-Staining Polypore

The Black-Staining Polypore (*Meripilus giganteus*) has larger lobes than Hen-of-the-Woods. The pale yellowish white undersurface bruises dark brown to black and the texture is very fibrous and stringy compared with the tender Hen-of-the-Woods. Although some people eat fresh specimens of the Black-Staining Polypore, severe digestive distress has been reported by some who have eaten this potentially indigestible ton of fiber.

Split apart a Black-Staining Polypore and you'll see stringy fibers.

The pore-covered undersurface of the Black-Staining Polypore bruises dark brown to black, usually within 20 minutes after handling. Hen-of-the-Woods does not discolor noticeably when bruised. Compare also against *Bondarzewia berkeleyi* (p. 43).

INTO THE FOREST

The Chicken Mushrooms
Laetiporus cincinnatus and
Laetiporus sulphureus

When mushroom hunters speak of the Chicken Mushroom—so named for its meaty texture as well as its mushroomy chicken flavor—they're talking about a big, wood-loving fungus that truly might feed twenty or thirty people. The Chicken Mushroom (sometimes called Chicken-of-the-Woods) can weigh 10 pounds or more.

This colorful and easy-to-identify fungus actually represents two distinct species in Illinois: one species has extremely tiny white or cream-colored pores on the undersurface (*Laetiporus cincinnatus*); the other has equally tiny yellow pores on the undersurface (*Laetiporus sulphureus*) and is also known as the Sulphur Shelf. There are additional differences, such as locations where they grow and color variations on the upper surface.

Laetiporus cincinnatus.

But whichever species you discover, know that both are wildly popular edibles, despite the troubling fact that some people suffer allergic reactions after eating the Chicken Mushroom. Vegetarians love it for its ability to convincingly mimic chicken in almost any dish, and everybody else loves that it can be found from April through October in Illinois, if only once in a while. But once in a while is really plenty of Chicken Mushroom.

Laetiporus sulphureus.

The Chicken Mushroom with the Yellow Underside
Laetiporus sulphureus

The eye-catching hue of this yellow-orange fungus can be spotted from a long distance, making them vulnerable to anyone who happens to recognize this delicious edible. Unfortunately, this species of Chicken Mushroom (also called the Sulphur Shelf) doesn't always agree with people after they've cooked and eaten this meaty polypore (see Comments).

Some people call it Chicken-of-the-Woods, not to be confused with *Grifola frondosa*, commonly called Hen-of-the-Woods, which is an entirely different mushroom (see p. 35). From late April through October, watch for living or dead hardwoods to erupt with massive fruitings of these overlapping, shelflike lobes. Good, fresh specimens should be juicy and soft—at least on the outer margins. This durable, wood-inhabiting mushroom lasts for weeks or months outdoors before turning crumbly, and for weeks they'll trick you into believing they're fresh, even when they're not, because their bright colors persist. Stacking them in open arms, you might find 15 pounds or more on one log or tree, only to discover once back in your kitchen that you really should have discovered them two or even three weeks ago.

The color of the undersurface, which will be either yellow or white, also shows on the edge of the upper surface.

The Chicken Mushroom with a white undersurface (*Laetiporus cincinnatus*) typically grows at the base of oaks, often a few feet away on soil above the tree's roots. Its fruiting habit is similar to Hen-of-the-Woods (p. 35), which never grows directly on wood.

The Chicken Mushroom with a yellow undersurface (*Laetiporus sulphureus*) always grows on wood, either on the sides of trees or on logs.

INTO THE FOREST

LAETIPORUS SULPHUREUS FEATURES

FRUITING BODY: Overlapping, yellow orange brackets or shelflike lobes up to 12 inches across and roughly ½-inch thick, sprouting from wood, often stacked or crowded in clusters. The outer edge is typically fanned outward from the base where it attaches to the tree or log. A thick, lumpy connection attaches to the wood. The upper surface has zones or bands of varying shades of yellow and orange, and the undersurface is sulphur yellow and looks almost smooth, but is covered in extremely tiny pores. Young, immature nodules of growth are mostly yellow with an orange inner portion near the base.

FLESH: Meaty, firm but tender when fresh. Yellow when fresh, becoming pale. When it breaks apart like wet chalk or damp cardboard, it's too old to eat.

STEM: There is no central stem to speak of; the entire cluster of caps is attached to wood by a dense base.

HABITAT: Always on wood, including living and dead trees, particularly oak, beginning in late April in southern Illinois and continuing through mid-autumn statewide. Its appearance isn't always dependent on recent rainfall, as specimens sometimes fruit even in fairly dry weather, supported by the tiny, unseen "pipelines" of fungal threads that supply moisture to the fruiting bodies.

COMMENTS: This is one of the polypores that swallows anything in its path as it grows. If a stick, another plant, or any other object happens to be where a Chicken Mushroom intends to put forth its growth, those objects will become part of the mushroom. Such interferences do not deter the growth of the mushroom, which continues to grow beyond the object as if it were invisible. One might find a poison ivy plant immobilized by Chicken Mushroom, the ivy and the mushroom perfectly oblivious to each other. Some individuals experience unpleasant reactions after eating this mushroom, including nausea, cramps, and skin rashes. Nonetheless, it remains a popular edible—but, like all mushrooms, it must always be cooked thoroughly.

The Chicken Mushroom with the White Underside

Laetiporus cincinnatus

These colorful, massive bouquets of pale orange fungus remind us of nothing else we can think of except, of course, a Chicken Mushroom.

There are two different species of fungi called Chicken Mushrooms in Illinois, and the main differences are the color of the undersurface and the habitat where they grow. This species of Chicken Mushroom, with its white undersurface and habit of growing above roots beside the base of oaks, is easily mistaken at a glance for the other Chicken Mushroom, which has a yellow undersurface and grows directly on wood. This version, which usually offers more tender portions overall, is preferred by most people, especially when it's collected at the peak of velvet-soft freshness. Cut the meaty lobes into ½-inch-wide strips and cook them up any way you would cook pieces of actual chicken.

The standard question: "Does it taste like chicken?"

Answer: Real chicken doesn't taste this good.

LAETIPORUS CINCINNATUS FEATURES

CAP: Fruiting body is a collection of shelflike lobes, up to 12 inches across, usually grouped as a cluster that can be basketball size or larger. Upper surface is light orange to nearly salmon pink, with alternating bands or ringlike zones of orange, pale yellow, or white extending from the outer edge inward; velvety smooth but lightly wrinkled with a peculiar tendency to have sticks and leaves poking out from within (see Comments). Undersurface is white to cream—not yellow—and covered with very tiny pores, not gills. Individual caps are shelflike lobes and connected in random association with other caps, arising from a central base in a massive bouquet.

FLESH: Meaty, but exceptionally tender when fresh, whitish cream to yellowish cream.

STEM: The entire cluster of caps is attached to the soil by a dense base, like a head of cabbage resting on the soil.

Young specimens of *Laetiporus cincinnatus* sometimes appear to have a faint, salmon-colored appearance on the upper surface instead of a bright orange color. *Laetiporus cincinnatus* typically grows at the base of oaks, or a few feet away from the trunk, where it appears on soil above the tree roots.

HABITAT: Beginning in May in southern Illinois until September statewide, they are found at the base of living or dead oaks and stumps, usually directly beside the tree or a few feet away on the ground where it appears above buried roots. Rarely found growing directly on wood—but always very close to wood.

COMMENTS: This is the preferred of the two species of Chicken Mushroom because nearly all of the mushroom is edible when young (its counterpart, *L. sulphureus*, is tender and juicy when fresh, but not all of it; usually only the outer edges of the caps of that species are worthwhile). Also, there seem to be fewer reported cases of allergic reactions to *L. cincinnatus* compared with its yellow-underside counterpart. Common problems reported include flushed face or temporary rashes, mild to severe nausea, and swollen lips and other alarming if not life-threatening reactions. For years this was considered a perfectly safe mushroom for anyone, which really isn't the case. Some individuals can't eat Chicken Mushrooms. Even if your internal chemistry doesn't tolerate it, the outward growth of the Chicken Mushroom is a marvelous, ghostlike phenomenon to observe because anything in its path becomes part of the mushroom. A blade of grass, a stick, an old piece of barbed wire—any small object becomes hopelessly embedded within the mushroom as it grows, immobilized like an insect trapped in ancient amber. The comparison isn't altogether far-fetched. Long-legged spiders, presumably waiting for prey or simply resting motionless for hours, have been observed with feet stuck in the growing flesh.

Mature specimens of *Laetiporus cincinnatus* tend to look a little more pale, although the orange color persists long after the mushroom becomes too old to eat.

CHICKEN MUSHROOM LOOK-ALIKES
Hapalopilus Croceus, Inonotus, Bondarzewia Berkeleyi

The seldom-seen *Hapalopilus croceus* resembles *Laetiporus sulphureus* from a distance. When young and fresh, the spongy tissue of this look-alike can be sliced easily (note penny inserted in top). With age, the bright colors fade to a drab, rusty brown. This woody polypore might remain on the side of oaks for months. Its edibility is unknown.

If this were *Laetiporus sulphureus*, it would be solid yellow on the bottom, not the top. The top would be orange, with a yellow zone or ring near the outer margin.

The interior flesh of a Chicken Mushroom does not have visible growth bands in profile.

Many different species of shelflike polypores exist in Illinois. Most are far too tough or woody to be edible. This is an *Inonotus*, a lumpy, inedible polypore with a fuzzy, velvetlike upper surface.

If this were a Chicken Mushroom—and it does resemble an old, faded *Laetiporus* at first glance—it would be much too old to eat anyway. This is a fresh *Bondarzewia berkeleyi*, a large, pale polypore found at the base of oak trees. Few people eat it.

Is It Still Fresh?

You'll eventually learn to recognize when a Chicken Mushroom is way too old to eat. But you'll get fooled many times beforehand until you figure out that this mushroom—like many other polypores—seems to last forever outdoors. The examples on this page, while attractive to foraging insects, are no longer fresh enough for human consumption.

People often spot a Chicken Mushroom from a distance, then hurry excitedly across the forest floor, only to encounter complete disappointment. They'll break off a slab of this still-colorful shelf fungus and notice that it snaps like damp chalk. That's when it's way too old to eat.

INTO THE FOREST

Pheasant Back
Polyporus squamosus

When you're picking morels, the world revolves around that one mushroom. That's why nobody bothers to pick the Pheasant Back, also known as Dryad's Saddle, a perfectly edible mushroom whose primary misfortune is its abundant availability during spring, exactly when morel hunters experience tunnel vision.

The Pheasant Back isn't rare. In fact, large brackets of this polypore commonly sprout from dead American Elm trees—the very place morel hunters will be crouched in April, picking morels. But everybody ignores this mushroom because they don't know anything about it. It bears a scaly, superficial resemblance to the backside of an Asian pheasant, hence the common name. It can appear in the fall as well as summer and spring. Also, if you cut it open and sniff, the aroma will remind you of the outer white rind inside a watermelon. It's such a big mushroom to let go to waste, and many educated mushroom hunters do indeed take it home, cook it up, and sample it—and it doesn't taste bad. But, for whatever reason, there are still a lot of Pheasant Backs left out there.

PHEASANT BACK FEATURES

FRUITING BODY: Typically large, shelflike bracket sprouting from both living and dead wood, logs, and so on. Beginning as a bulbous lump on wood, with a small dark brown to rusty brown top, the mushroom swells out to form a shelflike bracket, usually leaving a depressed area at the center of where it attaches to wood. A few days after rain, a puddle of water might remain in the cuplike depression. The irregular rows of brown scales are key traits of this polypore.

PORE SURFACE: Creamy white to yellowish white; pores very dense and nearly closed tight when young (creating the appearance of a smooth undersurface). In maturity, angular pores open and might even offer clear droplets of exudate, resembling dew.

FLESH: Soft, meaty, yellowish white. It smells like white watermelon rind.

STEM: Not much more than a short, thick trunk connects the cap to the wood.

HABITAT: On wood, both living and dead. Nearly any deciduous tree species can be a host, but American Elm is a major producer. Although it's commonly found during morel season, additional fruiting bodies can appear nearly anytime between spring and fall.

COMMENTS: This durable mushroom lasts a long time outdoors, but isn't worth collecting for food after it gets tough and leathery. Unlike some polypores that last for years on a tree, adding new growth each year, this one is shorter-lived—it eventually rots away a few weeks or months after it appears, just in time for another to appear. There's basically no other polypore found on wood that resembles the Pheasant Back. You'll recognize it. You've seen it.

When young, before the Pheasant Back fully expands into a shelflike slab, the darker cap has pronounced scales that stretch and flatten with age. White pores cover the underside.

Dead American Elm trees are common hosts for multiple fruitings of the Pheasant Back. Morel mushroom hunters know dead elms for other reasons.

INTO THE FOREST

Oyster Mushroom
Pleurotus ostreatus

Oyster Mushrooms are great to eat, and here's the best part. Oyster Mushrooms grow everywhere, anyplace there's wood, and nearly all year. Wherever you live in Illinois, you can be sure that this common mushroom grows someplace nearby—maybe right now. Therefore, all you need to do is learn where to look and you've got a continual supply of this mushroom-hunting staple. The name comes from the fact the caps often have an oyster-shell shape, not flavor. Still, they're great to eat.

Oyster Mushrooms appear frequently in Illinois, but they have no formal schedule. You might expect to find Oyster Mushrooms a few days after a perfect rain, and maybe you will—or you won't find any. They show up unexpectedly, even in icy weather, yet we're not surprised when it happens.

The trouble is, they sometimes seem to know how tall you are. Oyster Mushrooms can grow slightly above where your hand touches the tree while doing your very best unbalanced tiptoe reach, hopping like a little dog. Defiantly, people use sticks to pry them loose or climb trees—even as they know they are not at the age anymore for climbing trees.

OYSTER MUSHROOM FEATURES

CAP: Size ranges from medium to large (3–12 inches across). Oyster-shell shape is typical; surface smooth. White to silver gray in warm weather, often darker tan to brown in cool weather. Skin on cap can be peeled somewhat, especially when rain-soaked. Smooth, but sometimes incorporating bits of organic debris, bark, etc. on the surface.

GILLS: Thin, white to cream colored, forked near edge of cap. Gills run downward onto minimal stem.

SPORE PRINT COLOR: White to light lilac, never brown or any dark color (see "How to Make a Spore Print," p. 16).

FLESH: White, firm, and slightly rubbery.

STEM: Almost nonexistent or slight; whatever stem exists is almost entirely covered by the gills, sometimes with light, fuzzy hairs near base. The stem attaches directly to the wood, and clusters will often split open dead bark as the mushroom caps rise outward.

HABITAT: On wood, both living and dead tree trunks and logs. Also occasionally on the ground above buried roots or stumps. Nearly any deciduous tree species can be a host.

COMMENTS: This is one of the most popular among the "other" edible wild mushrooms in Illinois, partly because of the ease with which it can be identified, and partly because Oyster Mushrooms are one of the few species of edible wild mushroom that can fruit all year—and do so abundantly. Mushroom hunters literally cannot haul home everything they find, as

Oyster Mushrooms grow on both dead and living trees. Some examples might grow as large as dinner plates.

"Decurrent" gills beneath the cap of Oyster Mushrooms run downward onto a minimal stem.

INTO THE FOREST

tragic as that might sound. Oyster Mushrooms are called "Elephant Ears" by some people because of the shape of the large, mature cap. They can appear as a single cluster on one tree and nowhere else, or they can sprout everywhere, seemingly from every other tree and log in the right habitat. Willows near water can be outstanding producers—autumn waterfowl hunters know this edible mushroom well.

Oyster Mushrooms are easy to find, and that's a crowd-pleasing quality. People who collect wild mushrooms love it because it's a reliable resident of the woodlands of Illinois. If you take the time to look, you will certainly find Oyster Mushrooms at least once in a while, year after year, and you too will grow to appreciate this omnipresent fungus.

A true stem does not always develop on Oyster Mushrooms. But when it does, fuzzy white hairs that bruise easily might form. A similar edible species known as the Elm Oyster (*Hypsizygus tessulatus*) develops a fatter, bulblike stem, and that species usually appears alone or with one or two other fruiting bodies, and often on elm or box elder trees. Look for it fairly high on the tree trunk.

Light-colored summer varieties of Oyster Mushrooms often appear as individual caps.

Dark winter varieties of Oyster Mushrooms often appear as clusters. These specimens are very young.

Understanding Oyster Mushrooms
What's Typical?

Oyster Mushrooms, which always grow on wood, can fruit above the roots of dead trees, giving the impression they are growing from the soil. The relatively long stems shown below aren't typical because the mushrooms here rose from the soil above roots before forming a cap. The stems of most Oyster Mushrooms scarcely exist; they're mostly cap and gills with a short, offset stem for support.

What's typical? The scalloped-shaped lobes of the cap and "decurrent" gills running downward onto the stem are classic clues these are Oyster Mushrooms. Making a spore print (see "How to Make a Spore Print," p. 16) will reveal the diagnostic color of the spores, which should be a whitish lilac color—mostly white, with just a hint of lilac. Some wood-loving mushrooms with white gills (e.g., *Pluteus*, *Volvariella* species) produce pinkish to buff spore prints. All of those pretenders grow with proper, central stems and their gills are not decurrent. Notice also how the caps of Oyster Mushrooms never form a perfect circle around the offset stem.

Typical.

INTO THE FOREST

OYSTER MUSHROOM LOOK-ALIKES
Orange Mock Oyster and Bear Lentinus

Orange Mock Oyster (*Phyllotopsis nidulans*) has orange gills and a fuzzy, golden orange cap. An unpleasant odor is distinctive.

Both look-alikes have fairly small caps, typically no larger than 2–3 inches across.

Phyllotopsis nidulans.

Bear Lentinus (*Lentinellus ursinus*) has roughly serrated, toothed gills, a fuzzy cap, and a wickedly hot flavor when tasted raw. It's not edible. Spit it out.

Oyster Mushrooms can fruit above your grasp. Clever mushroom hunters extend their reach.

Oyster Mushrooms always grow on wood—but sometimes above roots or beside stumps, which creates the illusion they're growing from soil, which they're not. These dark Oyster Mushrooms grew above Chinese elm roots in southern Illinois near Carbondale. Cold weather often produces dark specimens. These were photographed in February. The same stump continually produced Oyster Mushrooms for years, with lighter specimens appearing in warm weather, darker specimens in cold weather (see "Understanding Oyster Mushrooms: What's Typical?," p. 49).

Several inedible species of small, gilled mushrooms with yellow to brown spores appear on wood (see "How to Make a Spore Print," p. 16). As a rule, be suspicious of "Oyster" mushrooms that don't have prominent, fairly large caps. Oyster Mushrooms can be small, such as the young ones shown here. But look-alike species (*Crepidotus* and *Panellus*, for example) never have caps larger than 1–4 inches across. Young, small Oyster Mushrooms tend to have wavy gills that straighten as the mushroom grows. Until you learn to recognize the Oyster Mushroom, collect only large specimens—5 inches across or larger—big enough to help rule out small, inedible species. Making a spore print (p. 16) will reveal the characteristic white-to-lilac spore color of Oyster Mushrooms.

In the Kitchen
with Oyster Mushrooms

You can now buy fresh Oyster Mushrooms in stores. This popular wild edible has been tamed for the masses, and it even can be purchased as grow-your-own kits, available in a showy assortment of colorful strains. When you buy Oyster Mushrooms, watch for freshness. Unlike the quick-selling Button Mushroom, a fair amount of stock is lost to spoilage with produce-aisle Oyster Mushrooms. The delicate, cultivated varieties quickly become slimy and rotten in those ever-wet produce displays, so shop very carefully.

Fresh Oyster Mushrooms should be firm, almost rubbery, when sliced. Discard limp, watery specimens—or don't buy them. Also, just because Oyster Mushrooms can be bought in a store doesn't diminish the culinary attraction of wild collections. Wild Oyster Mushrooms offer a range of flavor nuances, stronger or less so, depending on habitat and environment. The general consensus is that they do not really taste like oysters, but many chefs sometimes imagine they do and treat them as a mild, mushroomy ally in seafood dishes.

Any recipe that calls for "wild mushrooms" will work with Oyster Mushrooms. Slip them into any dish where regular mushrooms are appropriate, and you'll be glad you did. Reach for them as you would any basic ingredient. They're like common salt used to be, before the sea-salt craze.

Velvet Foot
Flammulina velutipes

This chilly-weather mushroom can be found growing on wood in December, which is a remarkable opportunity for mushroom hunters. Think of it: Fresh mushrooms can be found at Christmas.

Don't get too excited. The pleasure is rather small. The Velvet Foot—sometimes called the Winter Mushroom—doesn't have a lot of flavor. But this fuzzy-stemmed fungus isn't bad, either; when nothing else can be found during a hike in the woods, a handful or two of the Velvet Foot makes us feel good. The attraction of this edible species really is nothing more than that: It's edible and it's one of the few edible mushrooms found during winter. The flavor tends to be ordinary, by most appraisals. But they do make a fair pizza topping.

Be extremely careful with identification. There is a deadly look-alike, also found in winter, called *Galerina autumnalis* (see Comments).

VELVET FOOT FEATURES

CAP: Smooth, but tacky when wet; golden yellow orange when young, turning darker with age, nearly ruby red, especially toward the center of the cap. Often no wider than 1 inch across, almost never more than 2 inches across. Concave to dome-shaped when young, with a clean outer edge that should never show evidence that a membranelike veil once had been attached.

GILLS: White to cream, never brown or rusty. Somewhat widely spaced in proportion to the small cap, curving upward before attaching to the stem.

FLESH: White to cream, same color as gills. Agreeable "mushroomy" odor (if any) can be detected.

SPORE PRINT COLOR: White, never brown (see "How to Make a Spore Print," p. 16).

STEM: Dark brown to black near the base, lighter toward the cap, covered by fuzzy, velvet-like hairs similar to peach fuzz, especially lower. There should be absolutely no trace or

With age, the cap of the Velvet Foot turns reddish brown, darker toward the center.

When moist, cap is shiny and slightly tacky.

Gills are not tightly crowded together, and are white when young and dingy cream-yellow when old; spore color is white.

Attractive, two-toned stem color gradually changes from pale yellow above to nearly black below.

Velvety, fuzzy hairs cover most of the stem, becoming increasingly fuzzy near the darker base.

evidence there once had been a ring of tissue on the stem. The attractive, two-tone blend of light and dark color on the fuzzy stem should be quite noticeable, even in maturity. The stem is often short and thick (although cultivated versions of *Flammulina*, sold as Enokitake, or Enoke, feature long white stems and tiny white caps; it's a cultivation technique that renders them virtually unrecognizable as related to the wild Velvet Foot).

HABITAT: Most common in cool, even freezing, weather; appearing in clusters and as individual mushrooms sprouting from deciduous logs, stumps, and other dead wood, particularly elms, but sometimes also from wounds in living trees. It can appear side by side with Deadly Galerina mushrooms (p. 24), so be absolutely certain that each mushroom you collect is a Velvet Foot. Make a spore print.

COMMENTS: Collecting the Velvet Foot for the table really can be a life-or-death move, considering the consequences of misidentification. Although positive identification of this species can be easy, it's critically important to realize there exists a deadly look-alike for the Velvet Foot. Read all about the Deadly Galerina (p. 24) before trying this mushroom. Both mushrooms appear in cool weather, often on the same log or stump. Freezing and thawing throughout the winter causes both species to become somewhat darker and slimy, and the notable differences between the two will become less notable as the two species linger on through the cold. Collect only fresh specimens of the Velvet Foot. The caps are worth eating when cooked in a little butter. They do have a reasonably good mushroom flavor, but one must be absolutely positive with identification.

VELVET FOOT LOOK-ALIKE
Marasmius pyrocephalus

Marasmius pyrocephalus also has a fuzzy stem that tends to darken near the base. The differences: The long, threadlike stem is not a characteristic of *Flammulina velutipes*, and *Marasmius pyrocephalus* appears on the forest floor as individual mushrooms on fallen sticks. The Velvet Foot usually appears in clusters on trees or logs.

Wood Blewit
Lepista nuda

Just about the time of year when people unscrew the garden hose from the faucet, when there is no point in watering a garden anymore, that's when people wander over to look for a few wavy-capped mushrooms growing in their wood mulch. They are purple mushrooms, short and stout, and often with a fat base. These are Wood Blewits—right on schedule.

In Europe, people adore blewits for their fragrant aroma and fine flavor, and they're cultivated now and can be bought as grow-your-own kits, even though some people don't tolerate them well (see Comments). Still, there's no excuse for not setting down the garden hose and wandering over.

WOOD BLEWIT FEATURES

CAP: Medium to large caps overall, 2–6 inches across, concave to almost bell-shaped when young, smooth, but slightly tacky when wet, usually curving downward (almost abruptly) toward the outer edge—which is often curvaceously wavy. The color can be rich purple to faded lilac, tan, or even light brown, depending on age and weather.

GILLS: Purple to faded lilac; thin, fairly deep or broad, but curving upward before attaching to stem. Shorter gills extend from the cap edge inward, but not fully reaching the stem; odor is often fragrant, floral.

FLESH: Same color as cap, especially toward the exterior. It's as if the mushroom had been soaked in a violet dye, which has seeped inside the somewhat lighter interior, which is not quite white.

SPORE PRINT COLOR: Pinkish buff to faintly lilac—never rusty brown or orange (see "How to Make a Spore Print," p. 16).

STEM: Proportionately thick and stout, well-built to support the sometimes sprawling cap; often flared outward near base with a fibrous to slightly hairy exterior. The relatively short stem (commonly 2–3 inches tall), appears a bit lighter in color than the cap because of the lighter hairs that cover the purple surface.

HABITAT: Around wood chips, mulch, leaf piles, and so on. Found occasionally in cool weather as autumn sets in.

Wood Blewits' violet-colored gills curve upward before attaching to stem; compare with the violet, more widely spaced gills of *Laccaria ochropurpurea* (p. 73). There is never a trace of a ring on the stem; compare with toxic *Cortinarius* species (see Wood Blewit Look-Alike p. 58). Scent is pleasant; often described as fruity or perfumelike.

The edge of the cap of Wood Blewits is often wavy and creased downward slightly, especially with age. Don't be surprised to find weather-beaten specimens surviving late into fall. Wood mulch is a common host.

COMMENTS: What defines a Wood Blewit at a glance? The wavy cap margin and stout stature—along with some shade of violet in the gills, cap, and stem usually help mushroom hunters recognize the Wood Blewit from the distance of at least several feet. Closer inspection is always required, since mistakes can be made with similar-looking species of *Cortinarius* (see photo). The stem should be thick, often flared out at the base, and could fairly be compared to the strong trunk of a mighty oak, built for powerful support.

WOOD BLEWIT LOOK-ALIKE
Cortinarius

These toxic *Cortinarius* mushrooms resemble the Wood Blewit in their shape and lilac color. The differences: *Cortinarius* species have rusty brown spores (blewits have pinkish buff spores). Also, the gills of *Cortinarius* species are covered by a membrane when young. That membrane often leaves a trace of a cobweblike ring on the stem. Note how the rusty spores dust the gills and stem ring.

Toxic violet *Cortinarius* species have a cobweblike membrane covering the gills. At this young stage, the diagnostic rusty spores (a trait of all *Cortinarius* species) aren't yet mature and won't be visible.

Lion's Mane
Hericium erinaceus

From a distance, it appears as if somebody stuck a giant snowball on the side of a tree. That's *Hericium erinaceus*, which really does resemble a lump of snow. When examined closely, what appeared to be familiar becomes weirdly unfamiliar: Imagine tiny icicles dangling from a white blob. The common name is Lion's Mane, sometimes called Pom Pom du Blanc—a French term for those fluffy things cheerleaders wave around. Lion's Mane doesn't actually look like a lion's mane or cheerleader gear as much as it looks exactly like a lump of white stuff with spines.

It's pretty unique. Since nothing else really matches that description, you'll love encountering this easy-to-identify forest mushroom. People tend to find it right when the autumn leaves are spectacular or a bit sooner or later. It has a surprisingly mild, totally inoffensive tendency to actually taste a bit like crab or lobster meat, mild and sweetly tender, which is a culinary description rarely assigned to mushrooms. *Hericium erinaceus,* therefore, is a good example of why people should try a few different species of mushrooms in their life before deciding they don't like mushrooms. It's like being opposed to music based on a specific hatred of the zither.

LION'S MANE FEATURES

FRUITING BODY: Baseball- to basketball-size white mass of spongy tissue. Exterior "face" of the mass is covered with dense (but soft) spines that hang down like miniature white icicles. With age, or after heavy frosts, the white tissue might turn yellowish, especially at the tips. Its slightly rubbery tissue is often water-heavy, like a large wet sponge.

GILLS: None. Spores are produced on the white spines.

FLESH: Same color inside and out. Firm, slightly rubbery, like a big sea scallop. Mostly solid near the base, becoming hole-filled (like the so-called baby Swiss cheese) toward the outer edge as the branched network of "pipes" part company to form the little spines we see on the exterior.

STEM: None. Fruiting body is attached directly to wood.

HABITAT: One to several fruiting bodies on hardwood logs, stumps, etc. Also on tree wounds, usually in autumn. Not rare, but not common either.

COMMENTS: Two varieties of this basic form of *Hericium* exist: one has longer, thicker spines (up to 1 inch or more; the thickness of spaghetti); the other has shorter, thinner spines (usually less than 1 inch long and no thicker than a pencil lead.) The shorter variety has been reported from southern Illinois and Kentucky, and might occur elsewhere. Both varieties basically share all other traits and are equally edible. Distant look-alikes include a few hard, white polypores, some with burrlike spines. They might fool you from a distance. But nothing else has a spongy interior, can be easily sliced with a pocketknife, and essentially looks like slices of big scallop in the skillet. Cook it as you would similar seafood.

Hericium erinaceus subspecies *erinaceo-abietis*.

Bear's Head
Hericium americanum

Few things resemble the various mushrooms properly known as *Hericium*—except another *Hericium*. This species—*Hericium americanum*—differs from *Hericium erinaceus* in that the spines dangle from a loosely branched interior as opposed to a solid mass. It's more delicate than its meaty counterpart; therefore, it can't be sliced into neat, symmetrical portions. Most cooks work with its loose texture in the kitchen by serving cooked portions of it whole, or by combining it in dishes where odd shapes are welcome.

Hericium americanum seems to prefer rotted wood, a bit more advanced in decay than does *Hericium erinaceus*. Otherwise, look for both in similar woodland environments in the fall. Yet another species, *Hericium coralloides*, is equally edible but has shorter, almost bristly spines on its branches. It might be challenging for beginners to spot the difference. But confusing the two would be a harmless culinary mistake, since both are edible.

BEAR'S HEAD FEATURES

FRUITING BODY: Branched, white mass of soft spines growing from a central base on dead hardwood. Size varies from a small handful to an armful. When fresh, its attractive form actually resembles the branches of one of those Christmas trees that's been covered with white flocking, or the cascading icicles from a leaky fire hydrant in January. It basically differs from Coral Mushrooms (p. 62) in that countless white teeth or spines dangle—as opposed to growing upward—from its coral-like structure.

GILLS: None. Spores are produced on the white spines.

Featuring shorter spines on its branching, white form, *Hericium coralloides* is a closely related—and edible—*Hericium* species.

FLESH: Same color inside and out. Delicate, slightly rubbery, branching portions are a maze of spiny confusion.

STEM: None. Fruiting body is attached directly to wood.

HABITAT: One to several fruiting bodies on decaying hardwood logs, stumps, etc. Usually found in autumn. Found occasionally.

COMMENTS: This species of *Hericium* somewhat resembles coral, yet isn't one of the so-called Coral Mushrooms below. It has a similar, lightly sweet mushroom flavor as with other species of *Hericium*. But the loose, branching shape makes it a little less satisfying than the meatier *Hericium erinaceus*.

As the delicate spines shrivel and dry up, older specimens of *Hericium americanum* reveal the branched interior of the species.

BEAR'S HEAD LOOK-ALIKES
Coral Mushrooms and *Ceratiomyxa fruticulosa*

Many species of "coral" mushrooms exist in Illinois. None is known to be deadly poisonous, but most aren't worth eating. Some can cause stomach upset. The main difference between Coral Mushrooms and *Hericium*: Coral Mushrooms branch upward, while all *Hericium* species have soft, dangling spines.

The small, gelatinous tufts of inedible, coral-like slime mold known as *Ceratiomyxa fruticulosa* shouldn't be confused with the larger, downward-branching spines of a *Hericium* species.

Cauliflower Mushroom
Sparassis herbstii

When magicians tap a baton and a tissue-paper flower pops out, that's what the Cauliflower Mushroom looks like—at least, this cream-colored bouquet of wavy, thin vanes more closely resembles a tissue flower than it does cauliflower.

You'll recognize it when you see one. But that won't be very often, because the Cauliflower Mushroom isn't exactly a common mushroom in Illinois. However, you'll know when you see one because they can be as big as a soccer ball. Never plan to find a Cauliflower Mushroom on any particular walk in the woods. It just shows up once in a while in a lifetime, usually in late summer, in moist soils around conifers but also around hardwoods. It can be chewy, especially when old. But the flavor really is magical.

CAULIFLOWER MUSHROOM FEATURES

FRUITING BODY: Leafy cluster of pale yellow to creamy white ruffles, approximately 8–12 inches across, blossoming from a central base. Alternating bars of lighter and darker flesh—faint, lateral stripes—extend out to the thin edges. The cluster of ruffles is loosely packed, unlike the more dense (but also edible) *Sparassis crispa* of the western United States.

FLESH: When very young, the soft "leaves" break or tear about as easily as a tree leaf. With age, they toughen up (and get chewier when cooked).

STEM: Basically none. It has a central base, but the fruiting body is attached directly to soil, usually near wood.

HABITAT: In Illinois woodlands, it is often near cedars but sometimes without a clear association to any tree. It causes a brown rot of roots of hardwoods as well as conifers, although the mushrooms we see appear only infrequently overall, seemingly sprouting at random from the ground in moist woodlands. You might never find one in northern Illinois, and it's only slightly more common in southern Illinois woodlands. It tends to appear around August and early autumn, but only occasionally.

COMMENTS: Specks of dirt and forest debris sometime rise from the forest floor along with the leafy waves of this uncommon mushroom. Clean this mushroom as best as you can, then chop the leaves into strips and put them in a hot skillet with oil or butter. Few, if any, Cauliflower Mushroom recipes exist, so add them to whatever recipe might accept strips of thin but flavorful mushroom. You'll find it's surprisingly versatile once you accept its texture.

CAULIFLOWER MUSHROOM LOOK-ALIKE
Tremellodendron pallidum

These small, wavy clusters of white fungi known as *Tremellodendron pallidum* never grow larger than a few inches high. They're not little Cauliflower Mushrooms—and nobody eats them.

Wood Ear
Auricularia auricula-judae

Two important facts about this flabby-looking slice of rubbery fungus might surprise you: First, it's edible and, second, you've probably eaten it already.

Hot and sour soup, stir-fry mixes, and other Asian dishes often include strips of this basically tasteless but edible member of the Kingdom Fungi. A few species of this wrinkled mushroom exist around the world. This is the Illinois version. It goes by several common names (Judas' Ear being one of them), and it always grows from dead deciduous trees and branches, or dead wood within a living tree.

In dry weather the gelatin-filled tissue shrinks into tightly wrinkled, darker, tough and brittle, jerkylike *things*. You'll find them dehydrated for sale—usually Asian varieties, which are almost black—and they'll look plain wicked. The Wood Ear rehydrates just fine and becomes a perfectly edible mushroom you can share with dinner guests while serving Asian dishes. "Interesting texture" is how generous people describe the Wood Ear, before taking a drink.

WOOD EAR FEATURES

FLESH: Fruiting body is rusty tan to honey colored or motor-oil brown; somewhat translucent but not clear, wrinkled and wavy, basically disk-shaped when small—a floppy disk—growing into sloppy, irregular proportions. Individual "ears" can be about the size of a human ear but often larger, as much as 6 inches across. When very soft and fresh, the gelatinlike interior can be exposed by peeling apart the two outer surfaces. It's actually rather gross inside, and not much better outside.

STEM: None; fruiting body is attached directly to wood.

HABITAT: Individually or in clusters on a variety of deciduous trees, logs, branches, etc. Can be found almost year-round since fruiting bodies dry up but rehydrate themselves during wet weather. Most common in cool weather.

COMMENTS: This gooey-looking sheet of rubber has a long and ancient association with folk myth. Many years ago, somebody decided these earlike growths must be due to some curse on the tree or some spiritual bond between ears, trees, and the people willing to eat this peculiar fungus. These sometimes-prolific growths are best collected during moist weather, when the expanded shape of the mushroom easily proves its identity. When dry and tightly wrinkled, they might appear to be a bit of brown lichen, or a wrinkled piece of bark.

The flavor of *Auricularia auricula-judae* is not foul, nor is it good. We'd like to describe the flavor properly but it really has no flavor, as far as we can tell. Perhaps meditation will provide answers, or eliminate the question altogether.

WOOD EAR LOOK-ALIKE
Tremella mesenterica

Although more colorful than the Wood Ear, these orange, jellylike wrinkles, which turn brittle with age, are *Tremella mesenterica*, which nobody eats. Other varieties of jellylike fungi exist, including a clear-brown *Tremella foliacea*, which forms wrinkled clusters of jellylike muck on logs and sticks, and can be smeared when moist (unlike the Wood Ear). Also, a similarly gooey *Exidia* species commonly called Witch's Butter also can be smeared—unlike the rubbery Wood Ear. None of the look-alikes form large, fist-size lobes either. Until you learn to recognize the Wood Ear, collect only large, flabby specimens—or none at all.

Honey Mushroom
Armillaria mellea and Relatives

You are now looking at one of the world's largest living things.

This edible mushroom, the Honey Mushroom, grows out of the soil near infected trees and is typically 5 or 6 inches tall, maybe 7. However, the Honey Mushroom you see is part of the fungus *Armillaria*, which grows across the land, nourishing itself in tree roots and wood, sometimes surviving for hundreds to thousands of years. It can expand to staggering proportions—literally measured on maps.

First, scientists in Michigan found one in their soil: 37 acres around and at least 1,500 years old, they announced. The newspaper headlines were sensational: "Humungus Fungus in Michigan." The world of big records was smashed. Then, Oregon had one—2,400 years old, they said, and covering 3.4 square miles. Suddenly, every mycologist comparing DNA from *Armillaria* was digging soil plugs, taking measurements, searching for the next big thing.

Never mind all of that. Just be sure of your identification.

Armillaria mellea.

HONEY MUSHROOM FEATURES

CAP: 1–8 inches across, brown to honey-colored, especially when young; often fading to yellowish white with age. Lighter, hairlike scales or triangular specks might be visible. Sometimes sticky to the touch.

GILLS: Whitish cream, very thin, like tissue paper. Attached to the stem and sometimes tapering slightly downward on the stem. Shorter gills extend from the cap edge inward, but not fully reaching the stem.

SPORE PRINT COLOR: Spore deposit is whitish (see "How to Make a Spore Print," p. 16).

FLESH: White, very thin (a disappointment for the hungry mycophagist); it often seems the caps are nothing more than gills with a thin tissue covering the top.

STEM: Fibrous; one of the easiest stems to notice—but only after the cluster of Honey Mushrooms has been dead for a long while. The thin caps rot away quickly, but a trained eye will notice the cluster of

Armillaria tabescens.

black "spikes" lying on the forest floor. Those unrotted spikes are the lingering remains of Honey Mushroom stems. When fresh, the stems can have a somewhat iridescent sheen. That's because the tiny, fiberlike hairs that encase the softer, inner portions of the stem are somewhat light-reflective. Some species feature a ring around the stem; others do not. The stems usually seem fairly light in color, especially near the top. But just below the fibrillose exterior are shades of the namesake honey brown color.

HABITAT: Singly or in clusters around afflicted trees, roots, stumps, etc. Often appearing on lawns in clusters above roots. Silver maples in yards are a common host, as are oaks. Often abundant in autumn after rains, with several or more clusters sprouting in scattered bouquets.

COMMENTS: The Honey Mushroom is a good example of the old saying, "The more you learn, the less you know." That's because a lot of amateur mushroom hunters seem to know a lot about the Honey Mushroom. The problem is, nobody truly understands the genus *Armillaria* and all of its different varieties and species. Even the world's best mycologists have spent a lot of time peering into a microscope trying to sort out the different species of *Armillaria* believed to exist around the world. The Honey Mushroom "group" of related fungi are a confusing, overlapping collection of roughly similar fungi whose fruiting bodies sometimes are almost impossible to tell apart even with a microscope. Species have been separated based on their ability to mate with each other and based on DNA data. Some species are strong root pathogens while others are primarily decomposers of dead wood. There is no consensus as to how many different species of Honey Mushroom exist in the world.

Armillaria sp.

But, sometimes, just a little bit of knowledge really is a great thing. If you know nothing else about Honey Mushrooms, know that they must be cooked thoroughly. Severe gastrointestinal agony has been reported by some people after eating undercooked Honey Mushrooms. Also, individual reactions might vary, so nobody should eat too many on their first try. Most people discard the tough stem before cooking. Also, some Honey Mushrooms have a ring around the stem, while others do not. Some people seem to tolerate the varieties with a ring better than those without. But some people eat them both and can't understand what all of the fuss is about.

"Honey Mushroom" is a common name used to describe a few species of related fungi in Illinois. They can appear in seemingly endless physical variations—even within the same species—for reasons that aren't fully understood. Sometimes weather influences their appearance, including these rain-spattered specimens from central Illinois near Mattoon. The bell-shaped cap is unusual when compared with most other forms of *Armillaria*. The seemingly rough, scaly stem and cap are actually the result of grains of sand kicked up during a September thunderstorm. Compare with toxic *Pholiota* species (p. 70).

THE MANY FACES OF THE HONEY MUSHROOM

The underside edge of the cap that meets the cottony membrane or veil is not perfectly round, and can be a little wavy—a trait that persists through maturity. The edge of the mature cap is usually a bit wavy.

In all but one species of Honey Mushroom, a whitish, cottony veil covers the gills when young. As the cap matures, the veil breaks free but leaves a ring around the stem. Warning: Not every mushroom with a ring around the stem is edible—some are deadly poisonous.

Armillaria mellea.

Whitish silver stems are lighter toward the cap, but darken and show more honey brown color toward the base. Stems often are clustered in bunches, all arising from the same central point. Exception: *Armillaria gallica* (one of the common Illinois Honey Mushroom species) features a swollen stem base and often grows as separate, scattered mushrooms that do not arise from a central base.

The small, thimble- or helmet-shaped cap is often slimy when wet. The color ranges from darker to lighter.

Armillaria tabescens.

INTO THE FOREST

The caps of *Armillaria* often flare upward with age, and a dark spot or bump often persists at the center.

Stems usually exhibit an iridescent sheen because of the silky fibers on the outer surface. When examined closely, thin fibrous lines on the stem should be visible below the gills, running downward. Light, somewhat triangular-shaped tufts of silvery fibers on stem and cap form loose, scalelike designs. Similar marks are often noted on young caps.

Armillaria tabescens, a related species, has no ring around the stem, but roughly resembles *Armillaria mellea*. This ringless Honey Mushroom seems to cause more gastrointestinal problems—sometimes severe—and isn't recommended (see Comments). Most species of *Armillaria* have thin flesh in the caps—they're mostly gills covered by cap skin, without much meat.

HONEY MUSHROOM LOOK-ALIKE
Pholiota

A group of toxic, wood-inhabiting mushrooms representing the genus *Pholiota* sometimes fool Honey Mushroom hunters. One major difference: Honey Mushrooms always have white spores (see "How to Make a Spore Print," p. 16) and all *Pholiota* species have brown spores. Additionally, deadly *Galerina autumnalis* mushrooms have been seen growing among *Armillaria* mushrooms, so always look at every specimen to ensure that you haven't made a potentially fatal error.

Lobster Mushroom
Hypomyces lactifluorum

Most people can't stand to look at a Lobster Mushroom. Instead, they get on their knees and pull away the pine needles covering what they hope is a deliciously parasitized *Russula brevipes* or *Lactarius* species. This fluorescent orange mushroom often goes unnoticed because it barely pokes above the pine needles.

Of course, the hideously deformed appearance really is something, too. Lobster mushrooms are unappealing to most eyes.

But that's not what really bothers people. It's the fact that Lobster Mushrooms, like truffles, often grow in secrecy, hidden below the leaf litter. In this case, the habitat is just below a thin mat of pine needles or oak leaves, just enough to hide these crispy fungi from our view. But they can be found. And it's worth looking. Remember that truffles, those underground pearls of the oak, also win no beauty pageants.

Still, people look.

LOBSTER MUSHROOM FEATURES

CAP: Misshapen and distorted from the parasitic fungus that covers its surface. Instead of forming a typically shaped cap and stem, Lobster Mushrooms usually flare upward then outward, suggestive of a hideously misshapen chanterelle. The cap is usually depressed in the center, and is covered with tiny reddish orange bumps that make the mushroom appear to have been coated with barbecue potato-chip spice. The interior flesh is whitish to cream, with some seepage of the exterior color into the firm, nearly crispy or brittle flesh.

GILLS: None, because they were unable to develop beneath the parasitic *Hypomyces lactifluorum*. Occasionally there will be areas visible where it seems the gills tried to develop, but, in the end, just couldn't pull it off.

STEM: Connected to the cap seamlessly, with one becoming the other. Often funnel-shaped, sometimes split into lobes as the mushroom attempted to wrestle out of captivity. The whole thing can resemble a fat, reddish orange pancake that's been rolled into a funnel.

HABITAT: Technically, its habitat is on another mushroom, which will most likely be a *Russula* or *Lactarius* species. That species will be associated with trees in mixed forests, usually around conifers, sometimes hardwoods. Late summer to autumn is when they most often appear.

COMMENTS: This mushroom proves the adage that one's disfiguring misfortune can be another's delightful gain. In this case, the parasitic infection of *Hypomyces lactifluorum* suffered by the host *Russula* or *Lactarius* mushroom—mushrooms that ordinarily taste boringly plain—elevates this thing called the Lobster Mushroom to a truly delicious culinary treat.

Because the host mushroom is so distorted, it usually cannot be identified. Although we know of no reports of poisonings from eating the Lobster Mushroom (and it has been eaten for hundreds of years), there is a remote possibility that the Lobster you pick up may be parasitizing a poisonous mushroom. It is a small risk, but in our effort at full disclosure, we feel compelled to mention it.

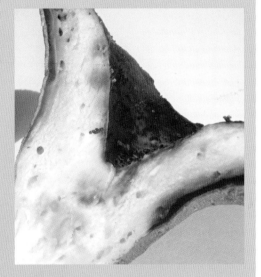

The fungus *Hypomyces lactifluorum* attacks the surface of *Russula* and *Lactarius* species, preventing the host mushroom from developing properly. Other species of green and yellow *Hypomyces* exist—avoid them.

Purple-Gilled Laccaria
Laccaria ochropurpurea

The Purple-Gilled Laccaria gives beginner mushroom hunters plenty of opportunity to find it in late summer and autumn. That's because this pillar of the mushroom community stands upright on the forest floor for weeks, seemingly still fresh enough to eat. But the Purple-Gilled Laccaria isn't a mushroom you'll want to ignore for weeks outdoors because the stem and cap seem to get chewier with age. When fresh, the whole mushroom remains quite firm when cooked and it holds up bravely to cooking. Unfortunately, the flavor isn't strong. This is one of those mushrooms you'll be glad to find outdoors when nothing else can be found.

PURPLE-GILLED LACCARIA FEATURES

CAP: Dry, convex to helmetlike when young, 1–4 inches across, often small in comparison to stem size, especially when immature; streaked with light brown to dark tan fibrils—a very drab, ordinary-looking cap when viewed from above. In maturity the top of the cap fades to mostly gray to light brown, with many specimens still seeming disproportionately small for such a sturdy stem. Thin edges of cap roll inward, often with an irregular, wavy shape.

GILLS: Very attractive, widely spaced, violet to purple—perhaps the most interesting feature of this mushroom. Gills are attached to the stem and very slightly decurrent (running downward slightly onto the stem), eventually losing their pretty color as they get coated with white spores in full maturity.

STEM: A hearty, strong support for a comparatively small cap, up to 6 inches tall in large specimens. Similar in color to the top of cap, often lightly violet or streaked with brown to tan fibrils, sometimes scaly, splitting roughly as the stem grows. It can be wavy or bent, thicker near the base, tapering somewhat toward the cap. When young the stem has the firmness of a radish or raw potato, becoming quite tough with age. This mushroom remains standing in the forest for days or weeks after maturity due to the indefatigable, fibrous stem.

SPORE PRINT COLOR: White (see "How to Make a Spore Print," p. 16).

HABITAT: Found around oak and beech trees, sometimes near pines or cedar. Usually found with several or more individual mushrooms growing in scattered colonies. Beginning in mid- to late summer in Illinois, continuing into the fall. Slow to develop, this mushroom might take a couple of weeks or more before slowly reaching its full size.

COMMENTS: One of the sturdiest "regular" mushrooms you will find in the woods, the Purple-Gilled Laccaria has a stem that is perfectly edible in addition to the cap, and that's good news because the cap can be comparatively small in proportion to the stem. The flavor is a little weak, but just good enough to make it worth collecting. It's not the only purple-gilled mushroom in the woods—yet no look-alikes feature this hefty stem. Cutting through the stem is like cutting through a radish. Most people identify this species at a glance by the combination of the strong stem and the beautiful, widely spaced purple gills. Purple species of toxic *Cortinarius* have brown spores and a cobweblike covering over the gills in young specimens.

Young specimens of the Purple-Gilled Laccaria can look out of proportion because of the small cap on the hearty stem.

The thin edge of the cap of the Purple-Gilled Laccaria curls over the gill edges, especially when young.

Indigo Milk Cap
Lactarius indigo

In Mexico, where wild mushrooms commonly show up in markets, this blue mushroom is one of those startling regulars known to Mexican mushroom eaters as *el hongo azul*. We suspect it has a better flavor in Mexico—it seems to cook up bitter here in Illinois. Regardless of our opinion, who could resist trying a blue mushroom that's perfectly edible when cooked? It's something to brag about at work tomorrow.

"Hey. Check it out—I ate a blue mushroom yesterday."

Lactarius indigo instantly bleeds brilliant blue liquid when cut. When fresh, it drips blue ink, and your fingers will be stained. This is a member of the genus *Lactarius*. All mushrooms in the genus exude some liquid when cut, usually milky or clear. This is the only mushroom that bleeds a strong blue when cut. With apologies to the color-blind, this one is ridiculously easy to identify.

INDIGO MILK CAP FEATURES

CAP: Medium-size, 2–4 inches across; light blue with darker blue rings extending outward from a center depression (the cap might form a primitive saucer that holds a puddle of water after rain), color fading with age to exhibit a greenish cast. When young, the cap margin curls downward over itself, but as the mushroom matures, the cap usually flares upward. There is a tendency for leaves and other forest debris to get embedded in the cap surface as the mushroom pushes up out of the ground. When fresh, the cap should exude an opaque blue liquid when cut. Drier specimens might not "bleed," but nonetheless still should show some blue ink.

GILLS: Light blue, thin, and connected to the stem. With age, the blue fades a bit, and a hint of green might show instead. Like all parts of this mushroom, blue ink should appear when the gills are scratched or cut.

SPORE PRINT COLOR: Creamy yellowish (see "How to Make a Spore Print," p. 16).

STEM: Light blue, rather short and nicely round. Interior is pithy, often becoming hollow in maturity. Blue ink appears when the somewhat granular stem is cut.

HABITAT: Grows on soil in mixed woodlands, found occasionally, beginning in early summer and continuing into early fall.

COMMENTS: The rather large group of mushrooms that "bleed" when cut are known as *Lactarius*, a genus that includes hundreds of species around the world. A few species of *Lactarius* can be pretty tasty, according to some people. But many of these latex-bleeding mushrooms are difficult to identify easily, and so we offer here only one species—perhaps the easiest *Lactarius* to identify anywhere. It's often said there are no blue foods, but here's an exception.

There are no "bleeding" look-alikes for *Lactarius indigo*. If you find a blue mushroom that bleeds blue, it's *Lactarius indigo*.

INDIGO MILK CAP LOOK-ALIKES
Cortinarius sp.

A number of different species of mushrooms found in Illinois feature some shade of blue or purple in their caps and/or stems, such as this *Cortinarius sp.* Some are toxic. The edible *Lactarius indigo* is the only mushroom that exudes a dark blue liquid when cut.

The gills of *Lactarius indigo* are blue, not white.

Attack of the Wild Mycophagists
Illinois Forests Are Filled with Competition for Edible Fungi

Research has shown that Eastern Box Turtles can live for forty years or more in the wilds of Illinois. Between those long winters spent hibernating underground, box turtles roam our forests eating mushrooms.

They eat different things too. Box turtles are somewhat omnivorous. But they sure do love mushrooms, especially mushrooms belonging to the genera *Lactarius* and *Russula*. Fortunately for humans, the majority of mushrooms that turtles eat aren't favored by humans.

Squirrels are equally eager to eat mushrooms. The same species of mushrooms turtles race toward are also grabbed by squirrels, then carried away to be munched on—like an acorn—from a tree branch overhead. Deer do eat mushrooms—if not morels—and ruffed grouse are reported to enjoy fungi as well. Some wildlife eat mushrooms that would kill a human. Never assume a mushroom is safe for human consumption simply because some other animal took a bite.

Eastern Box Turtle eating a species of *Lactarius* in Jackson County, Illinois.

4

The Morels

*Morchella esculenta, Morchella elata,
and Morchella semilibera*

Morel mushrooms are easily identified by their pitted, spongelike cap and hollow interior. Nothing else quite resembles a morel during spring—the season for morels. There are basically three distinct species of true morels in Illinois and all of them can be incredibly hard to spot among forest leaves, which adds to their legendary mystique. The flavor can be marvelous, making the hunt for morels a pursuit of culinary passion. Some mushroom hunters seem to be better than others at finding morels. It takes years of experience, good eyesight that's trained for the hunt, and humble patience. Being really lucky also helps.

Morchella esculenta, the Yellow Morel. This is the morel that makes morels famous. It can have a gray or yellow cap and can be as small as a child's thumb—or tall enough to make you blush. The pits of the caps are shaped like the holes in a sea sponge, with each pit resembling the wandering outline of an amoeba or bubbles in a lava lamp. The stem, creamy-white at first, eventually becomes brittle and granular, but tends to be stronger than stems of the other two species of morels. The flavor of this species, from central Illinois north, is wonderful—but weaker in the south. It's always hollow. A smaller variety (possibly a different species), sometimes found near yellow poplars in the south and mixed woods in the north, has larger pits for its size and is usually no larger than anyone's thumb.

Morchella elata, the Black Morel. This early morel is distinguished from other morels by its dark cap and pore ridges, which can be absolutely black, especially with age. Often the pores are aligned vertically, or roughly so. The cap might curve under itself before connecting to the stem—but the cap edge does not hang free as in the Half-Free Morel. The stem is creamy white-yellow when young and fresh, quickly becoming darker, somewhat translucent, granular, and easily broken. The flavor can be powerful, too strong for some people. It's a rare find in the Chicago area, but Black Morels first appear in southern Illinois in mid- to late March and can fruit continually throughout morel season, but mostly early.

Morchella semilibera, the Half-Free Morel. This simple morel is hollow, like all true morels. But the lower edge of the cap isn't connected to the stem—roughly half of the cap hangs free, and the pits are less chambered than other morels, with vertically aligned wrinkles—somewhat like the sea-sponge caps of other morels, but not quite. The crumbly yellow stem often grows quite tall in comparison with the little cap. The flavor is disappointing compared with other morels. You might collect just the the caps, if anything.

The Big Deal

Morels are the most popular wild mushroom in Illinois, hands down. People go crazy during morel season, and it's all because these hollow, sponge-capped oddities appear only once a year. Plus, they really do taste great. Also, morels pop up exactly at the time of year when everybody is craving to be outdoors anyway, soaking up the good, fresh greenery. Things are alive. Winter is history. It's the real new year.

Spotting a morel is a matter of luck at first, if you're all alone. People will take you morel hunting and they'll point at this or that area on the forest floor, but you simply will not be able to see what they see. They will be finding morels, and you won't.

"Here's one," they'll say. "Oooh— three. Four. Wow. Five."

Particularly annoying is when they walk across the woods to pick one at your feet, then point out another, the one you just stepped on. But then something that cannot be taught happens. You might blink, and then you start to *see* morels. Suddenly, you possess the ability to recognize those formerly invisible sponge caps on the forest floor. You are finding morels. Yes, these fabled

Tom and Vicky Nauman of Magnolia, Illinois, started Morel Mania—an international supplier of morel mushroom products—from their home in 1992. It's where the Illinois State Morel Hunting Championship was born in 1996.

mushrooms of spring are hard to spot, but at least you know what they look like now, and as quickly as you realize your power, the hunt becomes a desperate treasure hunt. You sense that all morels must be gathered as fast as they can be found before savage competitors find them.

But morels aren't so easily found. They're beyond difficult. You will continue to step hesitantly in the woods, unsure of your approach, hypothesizing how best to detect the undetectable—if only one could better see the almost-impossible-to-see, which is to finally recognize the transparent face of camouflage itself.

It can be done. But nobody can explain *how* it's done.

Once you taste morels, you will hunt for them every spring for the rest of your life. If you move away from morel country, you will mention morels every spring. If you stay, you'll get better at finding morels. With confidence, you'll gesture toward some wood lot while driving in December and say something like, "That looks good for morels next spring." But you will be guessing, honestly.

Morels make fools of experts, eventually. Yet all experienced morel hunters pretend to be experts, even as they creep ahead in the woods, staring at nothing and everything like everybody else. Every leaf is a suspect, every shadow a morel.

It really is a big deal.

Morel hunting in Makanda, Illinois.

What Else Eats Morels?
Do Wildlife Steal Our Mushrooms?

Deer browsing through succulent spring vegetation near Chicago's South Side didn't touch these morels, choosing instead the green shoots of young plants inches above the morels. Deer might be guilty of many things, but they are not known to eat our beloved morels. The same can be said of wild turkeys, the subject of a 2002 Southern Illinois University Cooperative Wildlife Research Laboratory study that examined the stomach contents of nearly 150 wild turkeys that hunters bagged during morel season: None contained any trace of morels.

Basically, the only animals known to regularly eat morels are humans—and little can be done to discourage *those* predators.

Deer in Cook County, Illinois, browsed the tops of these fresh plants but ignored the fresh mushrooms available directly below.

Verpa

The Morels That Aren't Really Morels

You'll be out collecting morels one day when you discover a curiously shaped morel that really doesn't look like a morel, although the yellowish stem suggests it might be related. You'll notice the edge of the helmetlike cap hangs freely over the stem—somewhat like the cap on a Half-Free Morel (p. 91).

What you've probably found is *Verpa conica*, one of the morel-season regulars a few people collect to eat. Unfortunately, some people have also gotten quite ill after eating a *Verpa*, so leave them alone. There are true morels waiting to be picked.

Verpa conica.

Verpa conica has a smooth, sometimes lightly wrinkled brown cap that hangs over the stem. The stem interior is filled with cottony tissue when young, but the tissue tends to fall apart when old or soggy. Still, the stem never appears fully hollow, as is the case with all true morels.

Top: a mature Yellow Morel (*Morchella esculenta*).

Middle: a mature Half-Free Morel (*Morchella semilibera*).

Bottom: *Verpa conica*; all from Jo Daviess County.

Yellow Morel
Morchella esculenta

When most people mention morels, this is the kind of morel they're talking about. It's commonly called the Yellow Morel, but it's also the same species people call the Gray Morel—or Blonde. The yellow, gray, and blonde varieties are genetically the same, like those different-shaped leaves on poison ivy.

Sometimes, gray morels become yellow as they mature, but not always.

Nature is filled with all sorts of tricks of individual diversity. Separation of the species requires a scientific tolerance for exceptions, and morel hunters face the same taxonomic puzzles mycologists face when sorting out all of the different mushroom species found in nature. For the sake of simplicity, this book lumps together all forms of the Yellow Morel. You'll soon recognize the common traits of the varieties—big, small, gray, tan, or yellow—which is all you really need to know.

YELLOW MOREL FEATURES

CAP: Size varies considerably, from finger-sized to as big as your fist—or bigger. Color ranges from gray to yellow-brown; characterized by honeycomb or spongelike pits. Upon close inspection, after splitting apart the morel to view the completely hollow interior, you'll realize the "cap" is simply the upper portion of the stem, covered with pits and chambers. Therefore, the cap and stem are essentially the same thing. A seamless connection delineates the spongelike "cap" from the barren "stem." It's a trait of both Yellow and Black morels. Individual pits are often arranged in a roughly vertical pattern, but also are irregular and amoeba-shaped. Never confuse the brainlike wrinkles and folds of false morels (p. 88) with the pits and chambers of a true morel cap.

STEM: Yellowish white, hollow, sometimes lightly wrinkled on the exterior, especially near base. Granular in texture, especially with age. During wet weather, the stem can grow rather tall, while the cap remains roughly the same size.

HABITAT: Beginning in early April in southern Illinois through early May in northern Illinois around dead American Elm trees, dead cottonwoods, old apple trees and other fruit trees, live ash and live Yellow Poplar trees (Tulip Poplar), and sometimes in fresh mulch no more than two years old. Also found in association with white pine.

COMMENTS: Like all true morels, the Yellow Morel is completely hollow inside. Nothing else you'll collect during the spring has a honeycomb cap and hollow interior (see "Morel Look-Alikes," pp. 84, 87, 88). Unfortunately, quite a few people cannot eat morel mushrooms because they experience gastrointestinal side effects. Always cook morels thoroughly. In a famous 1991 poisoning incident in Vancouver, British Columbia, seventy-seven people at a banquet became sick after eating marinated Yellow Morels—mushrooms that were not cooked.

Morels have completely hollow interiors.

MOREL LOOK-ALIKES
Stinkhorns and False Morels

Stinkhorn Morel, *Phallus hadriani*. In the fall, an eerily similar thing known as *Phallus impudicus* and *Phallus hadriani*—two of the stinkhorns—appear on the ground. These goo-covered morel look-alikes rise from a white or pinkish egg, and the honeycomb top is covered with a gooey brown slime. Stinkhorns attract insects with their foul-smelling slime, thereby tricking bugs into carrying away bits of spore-filled goo. Stinkhorns begin life inside egglike sacs just below the soil surface. Some people collect the eggs and cook them up as a culinary novelty before the putrid slime matures. Not to be outdone, some Chinese chefs chop off the offensive portion and add the clean portion to soups or stuff it like Cannelloni.

Another common Illinois stinkhorn, the Dog Stinkhorn (*Mutinus elegans*) rises from the ground, especially around mulch piles, and basically looks nothing like a morel. But it's a related stinkhorn, which might help you better understand the so-called Stinkhorn Morel.

The "Big Red" or "River Red," also known as *Gyromitra caroliniana* is closely related to a *Gyromitra* species which has killed people who've eaten it in Europe. Yet many generations of mushroom hunters in Illinois have eaten this mushroom without the slightest trouble of any kind. It contains highly toxic monomethylhydrazine, which dissipates during cooking. Although many people insist this is a perfectly safe mushroom, we cannot recommend it. The toxin builds up in the body—so even if someone safely eats False Morels over a period of years, they could be in for serious trouble.

The interior of *Gyromitra* species are packed with folds and dense tissue, making it relatively heavy compared with true morels, which are always completely hollow. Note also how the edge of the cap hangs free from the stem.

Brainlike folds and wrinkles are characteristic of the caps of False Morels. Plus, the edge of the cap hangs free from the stem. True morels have honeycomblike pits in their caps and (except for the Half-Free Morel) the edge of the cap connects seamlessly to the stem. In False Morels, the stem is not completely hollow.

The wrinkled lobes of *Gyromitra brunnea* don't match the honeycomb-type pits of true morels. A young specimen is pictured here.

Mature *Gyromitra brunnea*. Dirt and other debris are usually embedded in the tissue of the lower part of the thick stem. Compare with the stems of true morels, which might become dirt-spattered following a rain—but always can be brushed clean again.

Black Morel
Morchella elata

When somebody in Illinois shouts, "Morels are up!" it usually means somebody found a few Black Morels, which appear in the woods a couple of weeks prior to the Yellow Morel, on average. The essentials of a true morel are represented here: The cap and stem interior are always perfectly hollow, and the cap is pitted like a sea sponge or honeycomb. The flavor is magnificently intense.

Nobody has much luck predicting where Black Morels will grow. They are the most elusive of all Illinois morels because their color blends in so well with the forest—and they seem to have no predictable preference for habitat. Black Morels simply grow where they grow, but always near trees. You might find them in a dark ravine of a hardwood forest, or atop a sandstone cliff near a cedar—or next to a rusty chair in an overgrown trash heap. Nobody knows why.

BLACK MOREL FEATURES

CAP: Overall size of cap, not including the stem, might be 1–4 inches. Color is motor-oil brown when fresh, with darker—nearly black—edges on the pore ridges; honeycomb or spongelike pits often arranged in a roughly vertical pattern, but also irregular and amoeba-shaped. Hollow inside. Cap and stem walls are seamlessly connected, although the cap on many specimens has the tendency to hang over a bit before curving back inward to attach seamlessly to the stem; interior wall of hollow cap is connected directly to the hollow stem.

STEM: Yellowish white, hollow, sometimes lightly wrinkled, especially near base. Granular in texture, especially with age, when it becomes brittle.

HABITAT: Typically on rich soils from mid-March in deep southern Illinois to early May in the north. Found from southern Illinois north to Starved Rock State Park, but apparently rare in the Chicago area. Always associated with trees, but usually no specific trees—Black Morels have been found under hardwoods such as oak and hickory, but also beech, black cherry, and cedar as well as other trees. It doesn't hurt to look around Mayapple plants, which emerge from rich forest soils about the time Black Morels pop up. But the association is coincidental.

COMMENTS: Most mushroom hunters discover Black Morels by accident while they're out looking for Yellow Morels or while turkey hunting in the spring. Some people confuse them with old, way-too-mature Yellow Morels because the dark caps with the dark pore ridges seem to *look* old, even when they're still fresh enough to eat. Also, the tendency of the top of the aging cap to turn nearly black and pointed or constricted like a witch's hat suggests a rotten mushroom. Not necessarily; if you can still cut the stem with a knife without it crumbling away, it's probably still fresh enough to eat. But be careful—quite a few people have reported "allergic" problems after eating Black Morels—especially when consumed with alcohol. You should cook and sample just one Black Morel if you've never tried this species (a good rule of thumb for any edible wild mushroom), and then see what tomorrow brings.

Black Morels reappear in the same place year after year, often in the company of other morel species. Like other morels, they won't all pop up on the same day, but arise from the soil over the course of a few weeks. The rate of growth is dependent on soil temperature and rainfall. As the season progresses, as warmer days become the standard, Black Morels fade away while Yellow Morels persist.

Half-Free Morel
Morchella semilibera

The Half-Free Morel is the disappointing morel. It's mostly stem, with a little cap, half of which hangs over the stem (the origin of its common name), and the flavor is unsatisfyingly mild.

But it *is* a true morel—and it's perfectly edible, if you tolerate morels. The flavor can be decent, especially if dried and stored in airtight jars until some later date. In August, months after you've found fresh morels, open the jar and inhale deeply.

Not so bad after all, really.

HALF-FREE MOREL FEATURES

CAP: Usually no more than 1–1½ inches long, rather small compared with the caps of other morel species. Color is tan when young to motor-oil brown with age, usually with darker, blunt edges on the pore ridges. (Some people confuse its appearance with the Black Morel because of the overall color and dark pore ridges. See Comments.) Pits are usually open chambers arranged with vertical ridges, not so much like a honeycomb or sea sponge as with other morels. The edge of the cap hangs free over the side of the stem, connecting to the stem roughly halfway up the cap. In general, no matter how tall the stem grows, the cap remains the same size. With age, the fragile connection between cap and stem often breaks and a morel hunter might encounter a stem but no cap, and then find a small, thimble-size cap knocked to the ground.

STEM: Delicately brittle in comparison to other morels. Yellowish white, hollow, usually covered in almost powdery granules that are easily smeared away. In rainy weather, the stem can grow 10 inches or more, while the dismal little cap—where most of the flavor resides—remains not much bigger than a thimble.

HABITAT: Scattered in woodlands, often in the company of other morel species. Apparently more tolerant of excessive moisture and poor soils, Half-Free Morels can be found in low areas and new-growth forests. They often appear in the vicinity of Yellow and Black Morels, but not as a strict requirement.

COMMENTS: People often have trouble telling apart a Black Morel from a Half-Free Morel because the color of the caps have similarities. The differences become clear once you slice open the cap and stem and study how the cap connects to the stem: The edge of the cap on a Half-Free Morel hangs over the stem—totally free. Sometimes a Black Morel cap will mimic this design by curving under itself slightly before connecting to the stem. When in doubt, look for the delicate granules found on the fragile stem of the Half-Free Morel.

The hollow interior of the Half-Free Morel.

A Few Tips for Morel Hunting

Everybody wants to know where to find morels. It's the Big Question. But nobody ever reveals their morel secrets. You will improve your chances of finding morels every spring, however, if you look for certain kinds of habitat, and that will become easier each year with additional experience.

Yellow Morels—the most common of the three different morel species—favor the company of certain trees. Among all other tips anyone might offer, learning to recognize the several species of trees listed on p. 86 (under Habitat) will drastically improve your chances of finding Yellow Morels—because the sponge-capped treasures often grow around those trees. The rest really is up to you.

As you seek morel knowledge, ignore anyone who claims to know very much about morels. One might hunt for morels for eighty years and, in the end, realize they know nothing. Along the way, people will step forward with bold pronouncements. Scholarly claims of unlocked secrets will appear every few years. And while morels can now be cultivated in a controlled environment, nobody has much control over what happens in nature. And anyone who claims to know exactly what will happen next in nature really shouldn't be making such claims.

Morchella esculenta in Lake County, Illinois.

When Do Morels Appear?

Children waiting for Santa Claus to appear at Christmas have more patience than morel hunters waiting for morels to show up each spring. All morel hunters visit their morel patches far earlier in the season than they really should, hoping—through some unprecedented freak of nature—to find morels a week or two earlier than they did the previous year.

Morels basically grow at about the same time every year, a few days earlier or a few days later, depending on the whims of nature. But that's not what morel hunters want to hear.

Being the first person to discover that first morel of the season is a little like being the first student in class to finish a difficult exam. There is a tremendous amount of envious hatred directed toward the student who can stroll back slowly to his or her desk, the demonstration of important knowledge completed. Morel hunters dream of such glory.

And while the first morel is significant, for it shows that at least one morel exists in Illinois, it really is more pomp than anything else. Morels do not all show up overnight. One by one, like kernels of corn exploding with increasing rapidity, morels can be found throughout Illinois for a few frenetic weeks, beginning in late March in southern Illinois, until the popping subsides in early to mid-May in northern Illinois.

Climate change might ultimately affect the dates, but a person expecting to find Yellow Morels in southern Illinois should expect to find them on April 15. Around Chicago or Rockford, May 3 is a safe bet.

Black Morels appear in southern Illinois when the wildflowers known as Spring Beauties bloom.

Many morel hunters watch for what they believe are key indicators that coincide with the appearance of morels. When apple trees blossom, or when oak leaves are the size of a squirrel's ear, some morel hunters head to the woods expecting to find morels because they believe the time is right.

The trouble is, those folk beliefs that guide morel hunters into the woods are not consistently reliable indicators: Redbud trees might be in blossom, yet morels haven't yet appeared. Or Mayapple plants might rise from the forest floor—but morels haven't shown up yet. Various species of oak trees have leaves that develop at different times, and one species might have tiny leaves while another oak species has almost-mature leaves.

Soil temperature is fairly helpful, and many people find morels when the temperature of the soil in their morel patch reaches about 57 degrees. Keeping notes every year, with details about exactly what the various plants looked like during morel season, will eventually help you know with reasonable confidence when to expect to find morels. With each passing year, you will realize there is an amazing range of conditions during morel season, and you'll also discover that individual plants don't always show up at the same time every year. Plants that were in full bloom at the same time another plant was in full bloom one year might bloom a week or two apart the next year.

It will frustrate you. But it will also tease your curiosity, and every year, about two weeks before you think morels are scheduled to appear, you'll lose your battle with patience and walk out into the woods while knowing perfectly well morels should not be up yet.

But you will look anyway, because nature is hard to predict.

A knowledgeable morel hunter should be able to find fresh morels for approximately one month in Illinois. But many people continue picking and eating morels even after the mushrooms are dangerously too old to eat, like the one pictured here, increasing the risk of food poisoning.

Morels don't come with an expiration date stamped on a package. This ancient Yellow Morel is history.

THE MORELS

Jim and Shirley Nelson found their morel bonanza at Mississippi Palisades State Park in northwest Illinois.

One Sunday afternoon in April, Tammy Bryant heard that somebody had collected a bag of morels around Lake Shelbyville in Illinois. Bryant grabbed the kids, drove to the lake, and, just like that, found morels.

"Speedy" Elder of Union County, Illinois, had knee replacement surgery six months before morel season at nearby Giant City State Park. When morels popped up, he hiked for more than a mile to claim his share. It was a speedy recovery indeed.

Wisconsin farmer Jerry Smith found 33 pounds of morels on his property in one day during the Great Rainy Morel Season of 2006. Smith hauled bags of morels to a market in Muscoda—home of the Muscoda spring morel festival—and shared the wealth. In Illinois, morels can be bought and sold, but only those mushrooms that were collected from privately owned land. Mushrooms picked in Illinois state parks cannot be offered for sale.

Morels are where you find them. Mr. and Mrs. David Bushnell of Platteville, Wisconsin, had never found morels in their own yard before. Then, one day, they did. Out came the big bucket.

THE MORELS

About the time Mayapple blossoms appear in Illinois, morel hunters should hurry outdoors to collect the last fresh morels they're going to find until next year.

Yellow Morels often appear beneath live ash trees throughout the Midwest. Ash trees are easily identified by their netted bark pattern and symmetrical branches. Learning to identify the various trees where morels tend to appear improves your chances of finding morels.

5

The Chanterelles

Cantharellus cibarius, Cantharellus lateritius,
Cantharellus cinnabarinus, Craterellus cornucopioides,
and Craterellus foetidus

Chanterelles have a reputation bordering on the impossible. They can't possibly be as good as their reputation implies. Yet great chefs everywhere worship chanterelles—a truly wild mushroom that cannot yet be cultivated. People chatter with excitement whenever chanterelles are in season, which is all summer. Maybe it's because certain chanterelles are so easy to find. Their golden radiance gives them away from a long distance—plus, they can fruit in dazzling, massive colonies, like stars on the forest floor. When chanterelles are in season, you'll find plenty to eat.

Unless, that is, you're hoping for the totally impossible-to-see *Craterellus cornucopioides*—the Black Trumpet. Nobody ever sees Black Trumpets. One must first be led by the hand and shown their invisibility. But they're worth the hunt. Both authors rate the Black Trumpet as the best edible mushroom in Illinois.

Man, they're good.

Yellow Chanterelle
Cantharellus cibarius

Yellow Chanterelles are beloved by so many because they have a pleasantly delicate, buttery flavor that pairs supremely well with everything from scrambled eggs to broiled fish. When most people speak of "chanterelles," this is the mushroom of which they fondly speak. There are several similar-appearing species in North America; all of them are similar to the chanterelle of European fame. If you travel to a French market, you'll see chanterelles heaped in boxes delivered from the wild. They are exquisitely stylish. Celebrities eat chanterelles at parties—with artists.

The allure: Yellow Chanterelles possess a certain flavor that is quite good without being overpowering. Also, mushroom hunters love them because they're ridiculously easy to see on the forest floor. When Yellow Chanterelles are in season, you'll find them. None of this invisible-morel, eye-squinting nonsense.

YELLOW CHANTERELLE FEATURES

CAP: Yellowish orange to gold, fading to pale yellow, 1–3 inches across. When young, the cap is like an acorn cap—with the edge rolled over itself slightly. In maturity, the cap often flares upward like an umbrella in a windstorm, revealing the gill-like ridges below. Unlike most mushrooms featuring a stem and a cap, it's often difficult to decide on the exact point at which the cap and stem are separate. If one were to cut at the point below the cap where the gill-like wrinkles end, one would have a cap with what appears to be a portion of stem attached. For comparison, the cap of a store-bought Button Mushroom can be cut away from the stem with precision, separating the cap and stem perfectly, leaving no portion of either one attached to the other.

The flesh of the Yellow Chanterelle bruises golden brown, especially with age.

GILLS: A trademark detail of Yellow Chanterelles is the unique texture of the vague gills below the cap. They're not quite gills as much as they are mushroom corduroy—ribs of tissue rising and falling in narrow ridges and valleys under the cap. The ridges sometimes are forked, but smoothly so. Compare them with the thin, proper gills of the common Button Mushroom found in produce aisles. Those store-bought mushroom caps have blade-like gills, like the blade of a scalpel. That's a comparison one would never make with the ridges beneath a Yellow Chanterelle cap. When the mushroom is fresh, you should be able to smear the gills under a Yellow Chanterelle, as one might drag a fingernail through wax. Finally, the gills *always* run down the stem for a distance before fading away. Always.

STEM: Yellowish white, more yellow near the cap, usually with some white visible lower on the stem, if only at the bottom. As described above, the stem and cap aren't easily separated; the tissue comprising the stem is identical to the tissue comprising the cap, with one becoming the other. The stem and cap share mutual territory among the lower reaches of the gill-like ridges. More often than not, the stem can be spongy due to the burrowing of insects.

The Yellow Chanterelle's stem is light yellow but often whiter near the base.

FLESH: Odor can be fruity, but is not always distinct. White, firm when fresh, but also easily broken or crumbled. One could break off a piece of fresh Yellow Chanterelle as one might break off a piece of firm cheddar. With age, the mushroom gets a little tougher, slightly leathery, and the crumbly texture no longer exists.

HABITAT: Beginning in late May in southern Illinois to late June in northern Illinois around oaks—but wait for summer rains. As long as the weather stays warm, one might find a few in September.

COMMENTS: Many texts emphasize the strong, apricot-like aroma of Yellow Chanterelles as a diagnostic trait. However, collections made in some regions of Illinois often lack a strong apricot aroma. You might not be able to detect it at all. Farther north, in Wisconsin, the fruitlike smell has been observed in mature Yellow Chanterelles. Fortunately, Yellow Chanterelles make the job of positive identification easy because they often grow in endless quantities on the forest floor, like dandelions on a lawn. Abundance gives us the opportunity to examine the individual morphological variations among Yellow Chanterelles while emphasizing the important, shared traits.

When fresh, the gill-like ridges of Yellow Chanterelles can be smeared easily, as one might drag a fingernail through wax.

YELLOW CHANTERELLE LOOK-ALIKES

Certain other inedible species might roughly resemble the shape of *Cantharellus cibarius*, but key traits always differ.

If these were Yellow Chanterelles, wrinklelike gills would extend farther down each stem, as indicated by the dotted lines.

Hygrophorus pratensis cannot be Yellow Chanterelles because Yellow Chanterelles do not have thin, bladelike gills.

The poisonous Jack O'Lantern (p. 28) tricks some people into believing it's a Yellow Chanterelle because of the orange color and the gills that run downward onto the stem. But notice here the very thin, well-defined gills, which are not a trait of Yellow Chanterelles. The stems of Jack O'Lanterns are usually crooked and tapered at the base—and clustered together like a bouquet. Also, Jack O'Lanterns grow on wood, usually above buried roots. Chanterelles don't grow on wood; they grow on soil near trees. Although it's considered a fall mushroom, the Jack O'Lantern has been known to appear in May, about the time when summer chanterelles make their first appearance in southern Illinois.

Despite their poisonous status, Jack O'Lanterns can be a lot of fun. The gills of fresh specimens can glow an eerie greenish blue that is visible in the dark. Bioluminescence is known to occur in other species of fungi as well, including the mycelium of the Honey Mushroom (p. 67).

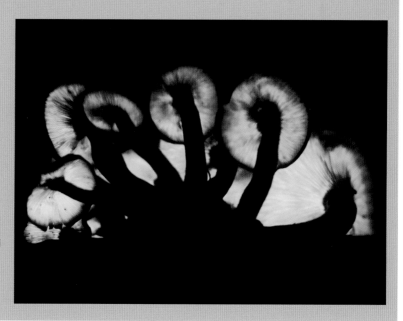

THE CHANTERELLES

Smooth Chanterelle
Cantharellus lateritius

Everyone gets fooled by Smooth Chanterelles, over and over again, thinking they've found prized Yellow Chanterelles. The trouble is, there's no way to positively identify either species from above. Mushroom hunters will be wandering through an oak forest in the summer, looking for chanterelles, when they encounter what appear to be Yellow Chanterelles scattered everywhere—and hearts will rise. But, like flipping a coin, the chances are only 50-50 they've actually got what they think they've got. One must look beneath the cap to verify the identity of either species. Whereas forked, ridgelike gills appear on Yellow Chanterelles, there basically are no gills on this mushroom.

It's a Smooth Chanterelle. It can exactly match the appearance of a Yellow Chanterelle from above and even grow side by side with its delicious relative. But looks aren't everything, especially among chanterelles.

SMOOTH CHANTERELLE FEATURES

CAP: 1–4 inches across, yellowish orange to gold, fading to pale yellow. When young, the wavy, thin edge of the cap curls down over itself. In maturity, the cap often flares upward but is usually uneven in its wavy appearance, and the center is depressed slightly. Surface is smooth. As with Yellow Chanterelles, it's often difficult to decide where the cap and stem become separate parts of the mushroom. They're seamlessly connected.

GILLS: None. The pinkish yellow spores are borne on a smooth undersurface beneath the cap—hence the common name. However, some specimens have faint wrinkles under the cap, almost as distinct as the corduroy-like gills of a Yellow Chanterelle. You will occasionally mistake one for the other, which is a harmless mistake between these edible mushrooms.

Cantharellus lateritius, Johnson County, Illinois.

STEM: Light yellow to nearly white, usually not as strongly colored as the cap, usually with some white visible lower on the stem, if only at the bottom. As described above, the stem and cap aren't easily separated; the tissue comprising the stem is identical to the tissue comprising the cap, with one becoming the other.

FLESH: White, firm when fresh, but also easily broken; usually more brittle than Yellow Chanterelles. Many texts describe the odor of Smooth Chanterelles as intensely fruity, as with apricots. But don't expect to easily detect the fruity aroma, especially in specimens from southern Illinois.

HABITAT: Beginning in late May in southern Illinois to late June in northern Illinois around oaks, following summer rains. As long as the weather stays warm, one might find a few in September.

COMMENTS: Positive identification is often simple because no other yellow orange, vase-shaped mushroom found around Illinois has a smooth undersurface. In southern Illinois, this species is more common than Yellow Chanterelles. Farther north, it's usually the opposite. People in Missouri once nominated this species as their official state mushroom not only because of its prolific abundance, but also because of its popularity for the table. It's a pretty mushroom, and it often grows to a larger size than Yellow Chanterelles, giving you more bulk per mushroom. The flavor can be decent to mild, depending on one's taste. In general, Smooth Chanterelles, like Half-Free Morels, are somewhat disappointing when compared with their delicious relatives.

Smooth Chanterelles might be perfectly smooth beneath the cap, or very faint wrinkles and ridges might appear instead. Many people have difficulty telling the difference between a Smooth Chanterelle and a Yellow Chanterelle, especially when the wrinkles are prominent on a Smooth Chanterelle. Both are edible. The smooth Chanterelle often grows larger than its counterpart and might feature a wavy, floppy-looking cap 4–6 inches across.

The edge of cap is thin and brittle.

Very faint wrinkles might be forked or netlike, or the undersurface might be perfectly smooth.

Wildlife often sample mushrooms (see "Attack of the Wild Mycophagists," p. 78), but not always because they like eating them. The Smooth Chanterelle at center has been ripped open by a creature that might have been searching for insects. For mushroom identification purposes, it's always helpful to recognize features caused by trauma or unusual growing conditions, as opposed to inherent traits of that species. The thin, wavy cap edge and the lack of gills are both legitimate traits of the Smooth Chanterelle—a shredded stem is not.

Catharellus lateritius, Jackson County, Illinois.

Red Chanterelle
Cantharellus cinnabarinus

Even after you've spotted a patch of these firecracker-red chanterelles—which are impossible to overlook despite their petite size—squint your eyes and look around, and look closely. There very well might be other kinds of chanterelles nearby—maybe Yellow Chanterelles, or Smooth Chanterelles, or even Black Trumpets. You might fill your basket with two or three different chanterelle species after spotting these simple little pleasures. Red Chanterelles aren't the most flavorful of the chanterelles. They basically taste okay; a bit peppery, most people say.

Some chefs ignore the always-cook-wild-mushrooms rule here by rinsing them and tossing a few into a salad. It's a colorful escape from salad boredom. One of the authors has eaten them raw without ill effects.

RED CHANTERELLE FEATURES

CAP: Fairly small, often no larger than 1 inch across (the entire mushroom might be no larger than 1 or 2 inches tall). Bright red to orange red. Smooth with a slightly irregular or wavy edge. When young, the relatively small, dome-shaped cap curls over the stem but soon flares upward into a vase shape, exposing the forked, wrinkled gills below. Caps are often small in comparison to stem size when young.

GILLS: Well-defined "gills" compared with other chanterelles. Netlike wrinkles might be present between each gill. The gills should run downward (decurrent) onto the stem, not merely stopping where the cap and stem meet. Gills are not crowded tightly together.

SPORES: As with all chanterelles, making a spore print from a Red Chanterelle can be difficult. The spore color is slightly pinkish cream, and can be seen by wrapping a few Red Chanterelles in clear plastic food wrap, then leaving them alone for a few hours. The spores will dust the inside of the plastic, which can be held over white paper to detect the creamy pink color.

STEM: Soft, rather fragile due to its small size but certainly strong enough to support the little mushroom cap. The reddish color is often streaked and of uneven density, like watercolors, with the lighter-colored, interior flesh of the stem showing through the thin, red "paint job," thus creating a somewhat orange cast in some areas. The gills should be visible on the upper portion of the stem.

HABITAT: Beginning in mid- to late May in southern Illinois to late June in northern Illinois around oaks, continuing into September when weather remains mild and wet. Usually found in patches or colonies of a few dozen or more, but with many of them simply too tiny and inconsequential to collect for eating. Although they usually appear in groups, the mushrooms do not grow in clusters arising from a central base. Each mushroom should be an individual, solitary growth, much like a group of dandelions are always comprised of individual plants.

COMMENTS: Red Chanterelles can be fairly easy to find because they often grow near other species of chanterelle, right in the same places you might be collecting Yellow Chanterelles or Black Trumpets. The quality of the flavor is debatable. Even when cooked, the flavor can be a little peppery. But a few people nonetheless find the flavor quite pleasing.

Red Chanterelles have well-defined gills.

RED CHANTERELLE LOOK-ALIKES
Hygrophoraceae and Mycena leaiana

The beautiful but inedible Scarlet Waxy Cap (*Hygrocybe coccinea*) has a sticky cap when wet and thin, well-defined gills.

Numerous brightly colored orange, red, and even green species of mushrooms commonly known as waxy caps belong to a collective group of fungi known as the *Hygrophoraceae*. Note the thin, well-defined gills.

Chanterelles never grow directly on wood. These orange *Mycena leaiana* are not edible.

Black Trumpet
Craterellus cornucopioides

Sometimes these brown-to-black funnels are no larger than a little toy trumpet on a child's doll. Other times they resemble a dark petunia blossom or a fairy-tale flower made out of thin, delicate chocolate. They're hauntingly somber yet naturally beautiful at once. Sadly, they're nearly impossible to see on the forest floor.

But don't give up. One day you will be hiking in an oak forest and find them scattered everywhere on mossy slopes, their cover of darkness betrayed by green. You'll see hundreds of them, clusters and more clusters. They'll be poking up everywhere, barely an inch or two tall, most of them. However, experienced eyes do have a better chance.

BLACK TRUMPET FEATURES

CAP: This brown-to-black, vase-shaped mushroom constructed of thin tissue doesn't have a proper cap. The entire mushroom is simply a hollow funnel, usually 1–3 inches tall, with the upper portion flaring outward, but sometimes forming distorted, club-shaped, hollow fruiting bodies. Upper surface might be flecked with a rough or minutely scaly texture. Outer edge might be black-rimmed, especially as the mushroom ages.

GILLS: None. Exterior, spore-bearing surface of trumpet-shaped mushroom is basically smooth, sometimes forming very faint, netlike wrinkles—but rarely prominent or distinct wrinkles as in *Craterellus foetidus* (p. 113).

SPORES: Salmon colored. Due to the shape of the mushroom, making a spore print is challenging with Black Trumpets, as is the case with all chanterelle species. But the dusting of salmon spores that often appears on the exterior surface of mature Black Trumpets helps reveal spore color (although the

Lucky mushroom hunters might fill baskets with Black Trumpets during the summer. It takes fairly good eyesight and patience to pick hundreds of little mushrooms. But very few pleasures can match the pastoral experience. Since Black Trumpets can be dried and then reconstituted perfectly, patient foragers stock up for winter when an abundant crop appears.

THE CHANTERELLES

combination of chocolate brown mushroom and salmon color often results in an orange cast.) Spiderwebs constructed near Black Trumpets can sometimes be helpful in catching spores and revealing their color. In the western United States, a nearly identical Black Trumpet has white spores. Until recently, the two were considered separate species: *Craterellus cornucopioides* was the Black Trumpet with white spores, while *Craterellus fallax* was the Black Trumpet with salmon spores. *C. cornucopioides* now applies to both versions.

STEM: Since the mushroom is funnel-shaped, the "cap" and "stem" are seamlessly connected as one. Black Trumpets sometimes rise from deep beneath leaf litter and therefore form long, tubelike stems before flaring outward into a cap. The stem is hollow but pinched closed at the base.

FLESH: Thin and soft, becoming a little tough with age.

HABITAT: Beginning in mid- to late May in southern Illinois to late June in northern Illinois around oaks. Black Trumpets can appear all summer, even into September if the weather remains mild. A preference for thin, exposed soils near rock outcrops is common in southern Illinois. Significant rainfall, not just one sprinkle or shower, seems helpful in triggering fruiting of massive colonies of Black Trumpets.

COMMENTS: Black Trumpets are amazingly difficult to see amid the camouflage of oak leaves on the forest floor. Most people don't notice them unless they happen to spot large colonies of this perfectly camouflaged mushroom. The flavor is very good when fried in a little butter, and it's one of the few mushrooms that can actually taste better in recipes after being dried and then reconstituted. Drying intensifies the flavor. Due to the thin flesh, the mushroom cannot be used in ways meatier mushrooms can be used. But in sauces, eggs, and casseroles, the Black Trumpet imparts a wonderfully rich flavor that compensates for its flimsy texture.

Black Trumpets are easily overlooked amid dead leaves in their oak forest habitat.

Look for Black Trumpets in exposed mossy areas where these delicate mushrooms retain their moisture.

BLACK TRUMPET LOOK-ALIKE
Devil's Urn

During morel season in the spring, mushroom hunters often encounter these dark, rubbery cup fungi commonly called the Devil's Urn. They're not Black Trumpets, and they're not edible. Black Trumpets typically appear weeks after these look-alikes have vanished from the woods.

Devil's Urn (*Urnula craterium*).

This look-alike is always found growing beside or above rotting oak logs and branches.

The edge of cap is ragged and the cuplike interior might hold rainwater for days.

Galiella rufa.

Black Trumpets do not have proper stems—they are funnels to the base. This inedible, stemmed, cup mushroom occasionally appears during Black Trumpet season.

These gelatin-filled cups always grow on wood. Black Trumpets grow on soil.

THE CHANTERELLES

Fragrant Black Trumpet
Craterellus foetidus

This species is closely related to the regular Black Trumpet but differs in two obvious ways: Whereas Black Trumpets are perfectly hollow funnels, the Fragrant Black Trumpet has a solid stem. Also, Black Trumpets tend to be smooth under the cap, maybe with extremely faint wrinkles. Fragrant Black Trumpets have prominently netted, forked wrinkles under the cap. You'll notice a hole or depression in the center of the cap, suggesting the stem is hollow, but it's not.

This very well might be the most fragrant edible mushroom known in Illinois. Regular Black Trumpets are wonderfully fragrant. But Fragrant Black Trumpets are absolutely exceptional.

FRAGRANT BLACK TRUMPET FEATURES

CAP: Gray brown to slightly orange, 1–2 inches across. As with most chanterelle-like mushrooms, the upper portion of this vase-shaped species doesn't form a proper cap. The entire mushroom is composed of a flared trumpet shape that doesn't form a separate cap and stem. However, the upper surface should have a hole or depression in the center. Its smooth to slightly scaly surface and wavy, sometimes black-rimmed edge are reminiscent of regular Black Trumpets.

UNDERSURFACE: Gray, a bit lighter than other parts of the mushroom. Instead of gills, netlike wrinkles extend downward onto the stem before fading away.

SPORES: Orange yellow to brownish orange. Making a spore print is challenging with any chanterelle due to their shape, and the Fragrant Black Trumpet is no exception. Try carefully wrapping a specimen in clear plastic wrap, waiting several hours, and then unwrapping the plastic to look at the spore dust (if any). Fortunately, spore color is not a major diagnostic trait for beginners attempting to identify this mushroom.

STEM: Gray to brownish, sometimes with a hint of orange—or, more precisely, the color of fresh motor oil—especially when compared with dark stems of regular Black Trumpets. This stem is solid below the cap opening, and somewhat brittle and fibrous. Interior flesh is lighter in color, almost cream.

FLESH: Thin and soft, becoming a little tough with age.

The Fragrant Black Trumpet has a hole or depression in the center of its upper surface.

HABITAT: Beginning in mid- to late May through summer in southern Illinois, around oaks. Found in similar habitat as regular Black Trumpets, but much less common.

COMMENTS: Fragrant Black Trumpets are every bit as difficult to see as regular Black Trumpets—but they're sure worth noticing. Since they're not hollow funnels like their featherweight counterpart, this mushroom is meatier and therefore more substantial in the skillet than Black Trumpets. Just don't expect to find vast carpets of this mushroom covering the forest floor, as one might see with regular Black Trumpets. You might hunt for mushrooms for years and never encounter this species due to its regionally uncommon status. But it is known to occur in Illinois.

FRAGRANT BLACK TRUMPET LOOK-ALIKES
Fluted Black Elfin Saddle

The somber-looking *Helvella lacunosa*, commonly known as the Fluted Black Elfin Saddle, is reportedly eaten by some people, but we don't recommend it. Very little information is known about the risks of widespread consumption of this species. Note the deeply ribbed, chambered stem and the thin cap tissue that bears superficial resemblance to the thin walls of a disfigured Black Trumpet. It can appear side by side with Black Trumpets and therefore is a risk for accidental collection by inattentive mushroom pickers. (Also see Black Trumpet Look-Alikes, p. 111.)

Hedgehog Mushroom
Hydnum repandum

One day you'll be in the woods, picking what you believe is a Yellow Chanterelle and you'll realize you've found something different. All over the underside of the cap there will be little white spines instead of gill-like ridges. Other than that, it really looks a lot like a Yellow Chanterelle.

You've found the Hedgehog, a bristly mushroom in a world of gilled mushrooms. They're not extremely common, but you might find a few. Their relationship to Yellow Chanterelles has been examined closely in recent years, and DNA analysis indicates the superficial resemblance we see is no coincidence: They're related.

From the standpoint of habitat, season, and physical texture, Hedgehogs should remind you of young, fresh *Cantharellus cibarius*, the Yellow Chanterelle—except Hedgehogs have spines under a proper cap. There are some other differences, also, and they're noted in Features. But here's what you want to know: If you like the flavor of Yellow Chanterelles, imagine them with a touch of walnut.

You'll honestly never find enough of these.

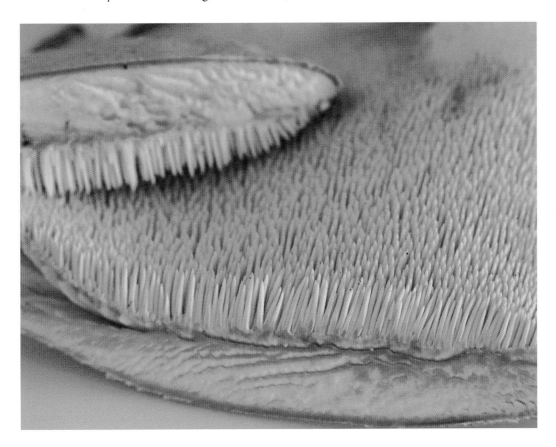

HEDGEHOG MUSHROOM FEATURES

CAP: Small to moderately large cap 1–5 inches across, yellow golden or tan, sometimes faded to pale cream. Upper surface dull, smooth, but not slick or shiny. Edge of cap curls under itself slightly, especially when young. Rarely perfectly round; usually irregular or with a slightly wavy cap edge. Very similar to the appearance of a young Yellow Chanterelle when seen from above.

UNDERSURFACE: White to cream-colored, needlelike spines of varying lengths cover the underside of the cap, although the spines sometimes appear yellowed or tan because of the discoloration that occurs at the tips of the spines. Take a closer look and you'll see the spines are, in fact, creamy white even if the tips are discolored. The spines make this an unmistakable find in the woods. The tiny, bristly needles, which extend downward slightly onto the stem, are the source of the common name for this mushroom. Spines should not be confused with pores, as seen on the underside of boletes (Chapter 6).

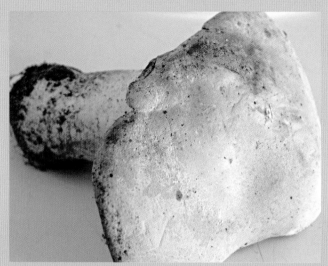

Hydnum repandum.

STEM: Smooth, white to cream-colored, bruising darker; thick, short, often no taller than 1 inch tall while supporting a comparatively large cap.

FLESH: Creamy white, slowly bruising to brownish yellow when bruised or cut. Odor can be nutlike.

HABITAT: Grows in soil around oaks; not as common as chanterelles but often found in the vicinity of chanterelles since their habitat is similar. Appearing in late May to early June in southern Illinois, later in the north. They can fruit into September in central and southern Illinois.

COMMENTS: A couple of versions of the hedgehog can be found in Illinois; one is typically no more than an inch across its cap and looks a lot like a small Yellow Chanterelle when viewed from above. The other has a

On the Hedgehog Mushroom, the edge of the cap curls down slightly when mature and more broadly when young, and is usually lobed or wavy, rarely round, especially when mature. The cream-colored stem slowly bruises darker, and the mushroom grows from soil, not wood.

more expansive cap, up to 4 or 5 inches across, with uneven, round lobes pushing outward, much like how pancake batter expands unevenly when poured into a skillet. This popular edible is consistently confused with Yellow Chanterelles at first glance from above. However, because of the diagnostic spines, you should confuse it with no other mushroom growing from soil (a handful of species growing on wood have spines). If your eyes aren't so good, you might have trouble at first glance telling the difference between the tightly clustered spines versus pores as found on such mushrooms as boletes (Chapter 6). Drag your finger across the undersurface—you'll figure it out.

Yellow Chanterelles (p. 100) and Hedgehog Mushrooms can look very similar from above, and they often grow together. But turn over the mushroom and you'll instantly see the difference. Yellow Chanterelles have gill-like folds and/or netlike wrinkles on the underside of the cap. The Hedgehog has crowded spines. Both are delicious, although you'll probably find a lot more Yellow Chanterelles than you will Hedgehogs in your lifetime.

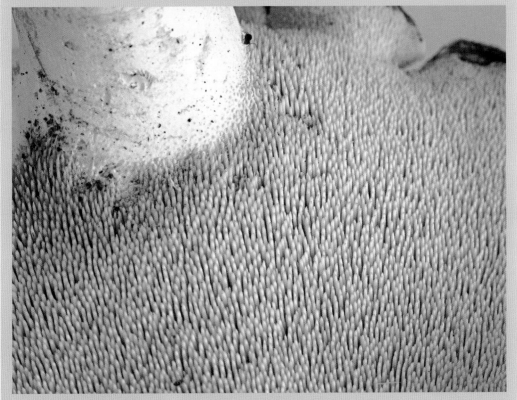

Spines of slightly different lengths crowd the underside of the cap of Hedgehog Mushrooms.

THE CHANTERELLES

6

The Boletes

Xanthaconium separans,
Strobilomyces spp., and
Gyroporus castaneus

There's really no point here in attempting to describe the often-tiny differences between the hundreds of unique species of mushrooms commonly known as boletes. This book is for beginners, so we'll start with basics: Boletes have a stem and a cap, just like the majority of mushrooms on the ground. The common feature among all boletes is that there are pores on the underside of the cap, not gills. Technically, those pores are really the ends of tiny tubes crowded together. But let's skip ahead to what's really important.

There are no deadly poisonous boletes known to exist. Some toxic species can make you violently ill, with lots of painful vomiting and diarrhea. But some boletes are quite edible and delicious. We'll be honest about this up front: There are so many damned different species of boletes out there, you'll wonder how anyone in the world tells them apart. We've selected a few that really are distinctive in unique ways, such as (hint) the mushroom pictured on page 119.

Chestnut Bolete
Gyroporus castaneus

There are more than 300 known species of boletes in North America, and they can be as common as they are difficult to identify, even for experts. Boletes, regardless of their scientific name, all have at least this much in common: Under the cap, instead of bladelike gills, boletes have pores that are actually the ends of tubes you'll see formed together under the "meat" of the cap. The amazing Cep of European fame is a bolete (*Boletus edulis*)—it's also called a Porcini. Knowledgeable Italians smile and shrug when saying the word, as if to add, "What else can I say?" People really worship *Boletus edulis*.

Unfortunately, Illinois has more polar bears than *Boletus edulis*. They just aren't found, or they're found just once every few years—by professional mycologists. But you will find a few Chestnut Boletes in your career, right here in Illinois. Once or twice every summer, you will go home, take the phone off the hook, and melt some real butter in a hot skillet. The rather small Chestnut Bolete isn't *Boletus edulis*. But it has a pretty good flavor. It really does. It could be a small Porcini if nobody told you otherwise.

CHESTNUT BOLETE FEATURES

CAP: Rarely much larger than 3 inches across. Dark chestnut brown when young (hence the common name). Like many boletes, the cap color of the Chestnut Bolete can be somewhat variable, but it will always have some shade of orange brown, sometimes a little reddish brown, sometimes a little yellowish brown. Perhaps it will be a little mottled, with uneven density in color, but never scaly or with fractures like cracked paint.

PORE SURFACE: The cap underside is covered with round, cream white pores that become yellowish with age—but not a bright, lemon yellow. If you have good eyesight, you'll notice the pores are basically round, which is a trait of all mushrooms within the genus *Gyroporus* (a Latinized scientific name meaning "round pores"). There should be no noticeable color change when the pore surface is scratched or bruised.

SPORE PRINT COLOR: Yellow (see "How to Make a Spore Print," p. 16).

FLESH: Like white marshmallow, but not sticky; dense. There should be no color change when cut.

STEM: Tan or buff with a bit of brownish color; usually swollen in the middle or at the base, tapering toward the cap; fairly smooth and without netlike wrinkles or veins. With age, the interior becomes somewhat hollow or pithy—and brittle—and is easily crushed when squeezed.

HABITAT: Mixed hardwood forest floors, but also near conifers. Beginning in June in southern Illinois, July in the north. Fruits into early fall.

COMMENTS: Nobody recognizes a chestnut bolete without first really

Chestnut Bolete, *Gyroporus castaneus*.

Pores beneath the cap of Chestnut Boletes tend to be round, although uneven expansion and growth can distort the shape. Compare with the often-angular shape of pores found on many other boletes.

THE BOLETES

looking at it, turning it over, and thinking. They're not hard to identify, but you have to look at a few typical features that should be present before announcing "Chestnut Bolete" and dropping it into your basket. They're tiny boletes. When fully grown, they're often no larger than an average, store-bought Button Mushroom—2 inches tall, or maybe 3. Identifying the Chestnut Bolete is accomplished by first picking up any small, brown-capped bolete that also has a fat stem toward its middle or base. The pores under the cap should be white if it's young; otherwise, the pores might be a bit yellowish. If the fat stem is somewhat hollow in addition to the other traits described above, it's a Chestnut Bolete. There are no deadly look-alikes matching this description. If you manage to find a good quantity of Chestnut Boletes someday, you might want to reexamine your collection to make sure you haven't carelessly tossed in a few vaguely similar boletes that really aren't Chestnut Boletes. There is no excuse for accidentally eating the wrong bolete, even if they all seem to look alike at first glance.

The brown color of the stem, the slightly swollen midsection of the stem, and nearly white pore surface of this small bolete should suggest to you that this is a Chestnut Bolete. Squeezing the lower part of the stem, which should be brittle and nearly hollow (slice it open to make sure), plus a yellow spore print, are all the clues you need to decide that this is a Chestnut Bolete.

The Chestnut Bolete, *Gyroporus castaneus,* which occurs in summer throughout Illinois, can be so small it seems hardly worth collecting. Notice the swollen stem (which should be somewhat hollow with age) and the chestnut brown cap. The light pore surface under the cap is additional evidence this is *Gyroporus castaneus*.

Old Man of the Woods
Strobilomyces floccopus

Sometimes people think all mushrooms look alike—but then there's the Old Man of the Woods. There simply aren't any other mushrooms out there that appear to have a cap covered with black-and-white tufts of soft wool.

The Old Man of the Woods (*Strobilomyces floccopus*) is one of the easiest to recognize of the Illinois boletes. You'll find it occasionally throughout oak woodlands from mid-summer into autumn. It's not extremely common. You might find one or two, here and there, following recent rains. Maybe five, if you're lucky. Fifteen would be a big haul for one day.

It's rather hard to confuse *Strobilomyces floccopus* with anything other than its twin species—*Strobilomyces confusus* (see Comments). They're both perfectly edible despite the black, inky juice that simmers out when you cook them. When dry weather sets in, this mushroom can remain standing for weeks, supported by its tough stem. Its cap might be totally unrecognizable as anything other than a charred marshmallow on a stick.

OLD MAN OF THE WOODS FEATURES

CAP: Medium-to-large cap, 3–6 inches across when mature, and covered with a thick mat of woolly fibers that splits into a mosaic of irregularly arranged, angular portions as the mushroom grows. When very young, the dense, convex cap is dark, nearly black. But that darkness splits open to reveal the white, woolly fibers as the cap expands. Eventually, everything on the cap turns dark brown, nearly black, which means the mushroom is basically too old to eat by then.

PORE SURFACE: The underside is covered with tiny, gray-white pores that might have a slight tint of salmon color due to the red-staining interior flesh. If there are any wounds or insect tunnels on the pore surface, they will be stained black because the pore surface bruises black after damage.

SPORE PRINT COLOR: Blackish brown (see "How to Make a Spore Print," p. 16).

FLESH: Like white marshmallow, but not sticky; dense. Immediately staining salmon red, then finally black, when cut.

STEM: Dark brown to black, with some of the soft, woolly tufts as seen on the cap, but not as exaggerated. Relatively slender when compared with some mushroom stems, but nonetheless sturdy, yet brittle. You should be able to snap it into pieces and possibly hear it snap. The strong stem can support a lot of weight for its size. Held aloft by the durable stem, the entire mushroom might linger upright for weeks on the forest floor, longer than other fleshy mushrooms. The stem is generally the same thickness from top to bottom, but slightly to ever-so-slightly thicker as it descends toward the base. You might notice a little iridescent sheen on portions not covered by the woolly scales. A remnant veil is sometimes present in some form—that's the raggedy shred of membrane that once covered the underside of the cap when the mushroom was very young.

Note the salmon red color change when the Old Man of the Woods is cut.

The distinctive cap of the Old Man of the Woods.

HABITAT: Mixed hardwood forest floors, often near oaks. Beginning in late May in southern Illinois, July in the north. Fruits into early fall.

COMMENTS: A look-alike species—*S. confusus*—is basically similar in all aspects except that the cap surface, which is usually soft and randomly shaggy on *S. floccopus*, is covered instead by pointed tufts of black wool on *S. confusus*. When quite young, either one could be mistaken for the other. But as long as every other trait matches up, you've found one or the other, and they're both edible.

The Old Man of the Woods, *Strobilomyces floccopus*.

Nothing resembles the woolly, black patches on the cap of Old Man of the Woods, making it one of the easiest boletes to identify.

Young specimens, such as this fairly large Old Man of the Woods from Shelby County, Illinois, often have dark, scaly caps. As the mushroom grows and expands, the fluffy, wool-like tissue pulls apart—almost like cotton candy.

Lilac Bolete
Xanthaconium separans

The next time you're buying laundry detergent, pick up a small bottle of ammonia, which will cost you perhaps a buck. With that small investment, you'll be able to convincingly identify this lilac-stemmed bolete for years to come, as long as the ammonia lasts—and it'll last for a long time. A mere drop on the pinkish stem of *Xanthaconium separans* should instantly change the color to a greenish blue or turquoise, depending on your eye and your familiarity with various hues of paint-store samples. It's the only bolete with a pinkish lilac stem that changes to turquoise with the application of ammonia. Even if you never wanted to be a chemist, you'll enjoy conducting this simple scientific test to confirm your identification. One bottle of ammonia—try to remember.

That stem is really what catches your attention first: Plenty of boletes have reddish stems. *Xanthaconium separans* is just about the only bolete you'll find with a lilac stem. "Raspberry blush" would be the name they use for the paint-store sample.

LILAC BOLETE FEATURES

CAP: Medium to large cap, 3–5 inches across. Dark chestnut brown when young, usually aging to a light brown or yellowish orange. The outer margin usually has a narrow, whisker-thin, white or yellow ring that borders the edge. The surface is typically lumpy with small depressions and soft wrinkles; otherwise, basically smooth, without scales or cracks.

PORE SURFACE: The underside of the cap is covered with tiny white pores that turn light yellow to lemon-yellow with age. There should be no discoloration when bruised. When sliced, you'll notice the crowds of tubes responsible for those tiny pores are somewhat longer than the thickness of the white flesh above—there are more tubes than flesh with this species.

The pinkish stems of Lilac Boletes.

SPORE PRINT COLOR: Olive brown (see "How to Make a Spore Print," p. 16).

FLESH: Soft, like firm white marshmallow, turning yellow with age; unchanging in color when cut. Not easily pulled away from the tube layer.

STEM: Pinkish lilac, but whiter near the base and/or near the cap. Typically there will be slightly raised, netlike "veins" on some portion of the stem. But those veins (called *reticu-*

The stems of Lilac Boletes become pithy or hollow with age.

The pore surface beneath the cap changes from white when young to lemon yellow with older Lilac Boletes. There should be no color change when the pore surface is bruised.

When young, the cap is often bowl-shaped and somewhat wrinkled or lumpy on the surface. The color is often dark chestnut ruby brown, but can appear lighter.

Netlike wrinkles or veins are usually present on some part of the stem.

THE BOLETES   127

lation) might not always be distinct, especially in specimens that have particularly fat or bulbous stems. In those cases, the stem surface often appears to be mostly smooth, usually with plenty of white. Sometimes the base will be swollen and/or hooked. The interior becomes fairly hollow or roughly chambered with age, resembling dried foam bubbles. A drop of ammonia on the stem instantly changes to some shade of turquoise.

HABITAT: Mixed hardwood forest floors. Early to mid-summer is most common, but perhaps you'll find it through August.

COMMENTS: It's always great to demonstrate an infallible scientific test to prove the mushroom you brought home is, in fact, the mushroom you thought you brought home. Any bolete that matches the traits described above should instantly react with a turquoise color change when a drop of ammonia is placed on the stem. If it doesn't, you've got the wrong mushroom. Don't fall victim to greed by collecting boletes with stems that maybe look like they could be pinkish if they weren't such a different color. Although there is a range of variation among the specific traits described here, it never hurts to test your identification with an inexpensive drop of ammonia. You'll discard the stem anyway, since it's often crumbly inside. Some people like to cut away the very soft tube layer on the underside of most boletes before cooking. The tubes can be rather mushy in a skillet, while the upper, white portion of the cap flesh remains firm. It's hard to be that snobbish with Lilac Boletes, since the cap has more tube layer than white flesh. Fortunately, the flavor of both is good.

LILAC BOLETE LOOK-ALIKES
Other Boletes

Boletus sp. The majority of bolete stems are solid, unless insects have tunneled inside. Lilac Bolete stems are often pithy with age, and seem almost hollow, especially near the base. A lot of boletes stain blue when bruised or cut, and people sometimes write messages on the flat pore surface for fun. Unfortunately, most of the boletes that stain blue aren't edible and should be avoided. The Lilac Bolete doesn't stain at all when bruised.

The cap and stem of Lilac Boletes should not be bright red like this *Boletus bicolor*. Many look-alike boletes have some red on the cap and/or stem. But when compared against the Lilac Bolete, the individual traits of potential look-alikes should rule out mistaken identity. Instead of yellow, the top and bottom of the stem should show some white.

7

The Puffballs

Calvatia gigantea, Lycoperdon pyriforme,
Lycoperdon perlatum, Calvatia cyathiformis,
and Calvatia craniformis

You've seen and kicked puffballs when they're really old: A puff of dusty spores billows out. Maybe you kicked twice, just for fun. What you're doing is what puffballs actually require because puffballs produce their spores on the inside, which means they depend on something or someone bumping into them to release their captive spores. Alas, that dependency doesn't always work out for puffballs, because sometimes people find puffballs when they're still young and fresh—perfectly white inside—and take those puffballs home and cook them up.

Giant Puffball
Calvatia gigantea

Somebody in New York once found a puffball more than 5 feet across, which is recognized as the world record. In Illinois, Giant Puffballs the size of volleyballs commonly appear on lawns and in woodlands in early autumn, but whoppers might reach the size an inflated automobile airbag. Giant Puffballs are reported from central Illinois north.

Identifying Giant Puffballs really is easy. Their size absolutely defines them. The real problem most people have with food this big is the same problem people would have with a giant zucchini or a gallon of pudding: It's impossible to enjoy it all without getting sick of it. Eventually, you will give up trying to find ways to cook and eat this monster mushroom, especially if the laxative effect some people experience after eating too many puffballs kicks in.

GIANT PUFFBALL FEATURES

EXTERIOR: A giant mass, sometimes softball size but usually much larger, up to 2 feet across. White to grayish and basically round, but often resembling a giant blob of rising bread dough. When examined closely, the surface resembles the surface of a bright full moon, with minor dimples or tiny cracks and a gray white, mottled appearance.

FLESH: Soft, like firm white marshmallow, but not sticky, turning greenish yellow with age, beginning at the bottom and then darker tan or brown as mature spores prepare themselves to be kicked to smithereens.

STEM: None. The bottom of the mushroom is connected to the soil with a slight "navel," and perhaps a minor, threadlike "root" or two. Removing a puffball should be no more difficult than lifting a ripe melon that easily falls away from the vine.

HABITAT: Common on lawns but also in forests beginning in August, continuing into October as weather chills. Often found at the edge of wooded areas at golf courses, forest preserve parking lots, and bike trails. Usually found from central Illinois north.

COMMENTS: The real challenge with this puffball is its staggering size. It's like one of those giant novelty cigars for which there is no practical use. After driving around town to show off this freak among fungi, you will want to return to your kitchen, wipe it off, then slice the puffball open to see if it's actually any good. If it's still white inside, it's still good. As with all edible puffballs, the interior must be totally white and free from the greenish-yellow color which appears with age and causes gastrointestinal trouble.

Few organic objects resemble a Giant Puffball. Still, the eye of the trained mushroom hunter is often fooled by anything white and round and not rolling on the ground.

THE PUFFBALLS

Pear-Shaped Puffball
Lycoperdon pyriforme

A hungry mouse or a small squirrel might be really satisfied to find one little Pear-Shaped Puffball in the autumn woods. Pear-Shaped Puffballs are disappointingly small—in fact, they're small enough to be almost inconsequential.

Fortunately, Pear-Shaped Puffballs rarely grow as solitary mushrooms. You'll find colonies of these marble-sized nuggets on logs and stumps after the weather turns cool, sprouting in crowded masses, bunched for us in easy-to-harvest clusters. You'll get fooled by some older ones, which will squirt out a puff of olive-colored spores when you try to pick them. Collect only firm, fresh puffballs with white interiors. A handful will be enough for supper, or two handfuls if you have company.

PEAR-SHAPED PUFFBALL FEATURES

EXTERIOR: Marble-sized, light brown to tan; surface speckled with tiny, bumpy spines that give a slight roughness to the otherwise smooth skin. Fruiting bodies are often crowded together, but individual examples are constricted near the base and might, with some imagination, actually resemble small, upside-down pears.

FLESH: White when fresh, turning dirty greenish yellow with age. When old and dry, little puffs of olive brown spores shoot through an opening in the top when squeezed. It's as much fun as popping plastic packing bubbles. But don't inhale the spores or you might get something called lycoperdenosis, a temporary, asthmalike condition that makes it difficult to breathe.

STEM: No proper stem exists. The base of the mushroom is connected to the rotted wood with white mycelial cords.

HABITAT: On rotted wood, especially during the fall, often fruiting multiple times on the same host throughout the season. You will find fresh ones beside old ones, and they might look alike at first until you pull them off and squeeze each one. Fresh puffballs are quite firm, and those are the only ones you should collect.

COMMENTS: Like popcorn, the small size of this edible treat is no problem because you'll grab handfuls of these pear-shaped mushrooms once you locate a productive log or stump. The small size is actually an asset in the kitchen: The individual balls can be halved or quartered and cooked like any other mushroom. Make sure they're pure white inside. Thanks to the thin skin that covers the softer white interior, these cooked puffballs aren't quite as limp or mushy as larger puffballs, and the flavor is fairly decent. Avoid eating any puffballs that have dark interiors. A number of other puffball-shaped fungi exist, including some toxic species. But none of the troublemakers is pure white inside; in fact, some are perfectly black when split open.

A slightly larger, golf-ball-size puffball known as *Lycoperdon perlatum* (p. 129) has a roughly spiny brown surface and somewhat resembles the Pear-Shaped Puffball except that it favors mulch and soil instead of rotting logs and tends to grow solitarily or in small groups of two or three puffballs. Like all puffballs with a pure white interior, *Lycoperdon perlatum* is edible, although you might want to peel away the burr-covered surface before cooking.

Purple-Spored Puffball
Calvatia cyathiformis

In southern Illinois, where *Calvatia gigantea*—the Giant Puffball—doesn't seem to grow, mushroom hunters instead find these big balls of light brown mushrooms growing on their front lawn and think they've really got something.

They do. The Purple-Spored Puffball—*Calvatia cyathiformis*—is every bit as tasty and edible as any other puffball out there. When it gets old, the spores turn purple and produce that proverbial purple haze when stomped. But when it's young and fresh and pure white on the inside, you can cook it up like the best puffball you ever found.

You'll find too many to eat when these softball-sized mushrooms are out, which is the same problem people in northern Illinois have when they find one Giant Puffball.

PURPLE-SPORED PUFFBALL FEATURES

EXTERIOR: Softball-sized, tan to buff, somewhat round but usually developing a plump base and a larger, swollen upper portion. The shape can be somewhat irregular, like rising bread dough.

FLESH: Solid white when fresh, turning dark burgundy purple with age. When old and dry, the upper, spore-filled portion can be stomped to release spores. But it eventually weathers away, leaving a dusty, purple base that can persist until next year's puffball season. Stomping on the weathered base produces almost no spore cloud, since the spores are formed in the upper portion, which disappeared long ago.

STEM: No proper stem exists. The smaller, sterile base of the mushroom (which persists for months) is connected to the soil with white mycelial cords.

HABITAT: On lawns, pastures, cemeteries, and other grassy areas in late summer, especially in southern Illinois.

The Brain Puffball (*Calvatia craniformis*) favors woodlands as opposed to the grassy areas where the Purple-Spored Puffball grows. Another difference is that this species has a yellowish brown interior spore mass when it's too old to eat. But when it's young and perfectly white inside, it's as good as any other edible puffball.

COMMENTS: Purple-Spored Puffballs can grow as big as a head of cabbage sometimes, nearly rivaling the massive size of the famous Giant Puffball of central and northern Illinois. But none of these southern Illinois "giants" ever gets as large as a basketball. Softball size is most common. The thick rind on the exterior can be peeled or cut away, leaving nothing but pure white mushroom for cooking. Some people slice them, bread them in seasoned crumbs, and then fry them like eggplant. People tire of them easily, like too much of anything.

With age, the purple spore mass of an old Purple-Spored Puffball held within the upper portion of the puffball disintegrates—spreading spores everywhere—leaving behind a lightweight base that might persist for months outdoors.

PUFFBALL LOOK-ALIKES
Amanitas, Galiella rufa, Scleroderma citrinum

Deadly poisonous *Amanita* mushrooms begin life encased in an egglike membrane partially buried in soil. Although it's improbable anyone could confuse a partially buried *Amanita* with an edible puffball (puffballs always grow above ground or on wood), never cook a puffball without first slicing it open to make sure the interior is pure white and without the undeveloped "embryo" of an immature *Amanita*.

The deadly poisonous *Amanita bisporigera* and its egglike sac.

Many kinds of round or disk-shaped fungi grow on wood, but none really matches the basic appearance of a puffball. These inedible jelly-filled cups, *Galiella rufa*, which start life as tiny brown balls on fallen twigs, are easily recognized by the brown gelatin that fills their interior.

Always slice open a puffball to check inside. Puffballs with dark interiors, like this example of *Scleroderma citrinum*, should never be eaten.

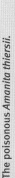
The poisonous *Amanita thiersii*.

Take the Field without Getting Hurt

Agaricus, Coprinus, Macrolepiota, and Lepiota

The first mushrooms we ever noticed as kids were probably on our lawn. But, even as children, we already knew not to pick those mushrooms. Since toddlerhood, we'd been taught that nobody should ever touch or eat a wild mushroom.

They could kill you.

"*Kill* you," Mother repeated. It's the first time she ever shook a finger at us. And so we listened. But the truth is this: Some mushrooms growing on your front lawn are perfectly edible, and quite delicious. It's true. But some of the mushrooms on your front lawn are also poisonous.

As a matter of fact, the mushroom responsible for most mushroom poisonings in America—*Chlorophyllum molybdites*—grows on lawns everywhere during warm weather, and looks harmless. It's not. Many have been poisoned.

Mothers everywhere will be pleased to know that their warnings were true.

In the Beginning: *Agaricus*

Agaricus. It sounds like something Caesar would proclaim to Romans, doesn't it? As in, "Hail valiant *Agaricus*! Hail!" Actually, Caesar really might have said something like that (although probably not in English) because *agaric* is the Latin word for mushroom. Caesar ate mushrooms; in fact, one actually bears his name: *Amanita caesaria*.

With age, the Meadow Mushroom's pink gills turn deep chocolate brown, the color of the spores. A cottony membrane covers gills when young. As the cap expands and breaks the membrane, a ragged collar or ring remains.

For a long time, all mushrooms were simply known as *agarics*, just like all mushrooms today are called *mushrooms*. And then there's *Agaricus*—the scientific name used to describe one of the first mushrooms ever to be given a scientific name, the same mushroom we now recognize from pizzas, steak toppings, and stuffed mushroom caps. *Agaricus* comes from the Latin word *agaric*, and *Agaricus bisporus* is the official scientific name given to describe today's ubiquitous Button Mushroom. But let's move on.

Perhaps the reason so many of us fear wild mushrooms is because we don't recognize a clear difference between those wild mushrooms growing on our front lawn and safe, produce-aisle mushrooms. But here's the deal: The reason that it's hard to tell the difference between mushrooms on your front lawn and produce-aisle mushrooms is because they might be very closely related—some of them anyway. Others can be poisonous. *Agaricus bisporus* is the produce-aisle mushroom. *Agaricus campestris* is the wild relative to the produce-aisle mushroom.

There are more than 100 species of *Agaricus* known to exist in the United States. Many of them are edible, and many of them are quite good, even better than commercially grown *Agaricus*, but some can inflict digestive trouble—serious digestive trouble, if you get the picture.

In Illinois, there are no deadly poisonous *Agaricus* species. However, some species growing in the woods or around trees are risky for the digestive tract. The authors suggest a general rule for safe *Agaricus* picking: If it has pink gills underneath the cap and has grass clippings on top, it's basically as safe as any store-bought mushroom—assuming, of course, the lawn hasn't been treated with toxic chemicals. People have suffered severe cases of herbicide poisoning from eating mushrooms collected on treated lawns.

The *Agaricus* species that favor grassy areas—far away from trees—are edible as long as they have pink gills when young. Nothing else on your front lawn, based on this rule, will trick you into believing it is edible when it is not.

Meadow Mushroom
Agaricus campestris

During the heat of summer (and until cool weather sets in), experienced mushroom hunters around the Midwest keep an eye out for something they call "Pink Bottoms." It's a fitting, alternative common name for the Meadow Mushroom because it helps describe a key feature of *Agaricus campestris*. When young and fresh, the gills of this species should be rosy pink. As the mushroom ages and spores mature, those pink gills turn dark chocolate brown due to the spore color. Realize that toxic look-alikes also can have brown gills, so learn to identify the Meadow Mushroom based on young specimens with pink gills.

Look for Meadow Mushrooms in mid- to late summer and into early autumn if the weather stays warm. Collect only those mushrooms found in grassy areas, away from trees, since some related *Agaricus* species (also with pink gills) are found around trees and those species could be toxic.

MEADOW MUSHROOM FEATURES

CAP: Medium-size, 2–4 inches across; white when young, becoming silvery white to off-white with age, sometimes very slightly tan or golden, as if baked in the sun. When young, the round white cap might seem like a golf ball in the grass. There should be very fine, soft, silklike fibers on the surface, increasing near the edge—not to be confused with the powdery, cottonlike fibers found on the caps and stems of the toxic *Amanita thiersii* (p. 30). *Amanita* gills are white or a bit yellowish with age. When the Meadow Mushroom cap is young, those cottony threads connect the edge of the cap and the stem, forming a protective covering for the gills. As the mushroom grows, the cap expands and flattens out and the protective veil tears apart, usually leaving a ragged ring or collar on the stem.

GILLS: Narrow, rosy pink when young, becoming dark chocolate brown with age, nearly black.

SPORE PRINT COLOR: Dark chocolate brown, as are spores of all *Agaricus* (see "How to Make a Spore Print," p. 16).

STEM: White to silvery, rather soft (like the stem of store-bought Button Mushrooms), slightly fibrous, fairly stocky at times and sometimes slightly thicker toward the base, especially in short specimens. There should be no evidence the mushroom rose from an egglike sac (universal veil) underground. Some species of *Agaricus* have a small bulb at the base. This one does not. The stem of *Agaricus campestris* can be quite soft and easily crushed when wet or after rain. It should bruise a bit darker and be translucent when handled or crushed, but should not stain yellow, especially at the base. Some possibly poisonous woodland *Agaricus* species bruise yellow at the base. Do not pick any *Agaricus* species in the woods or directly in the vicinity of tree roots. Remember: Meadow mushrooms grow in open, grassy areas.

HABITAT: Your front lawn, your neighbor's front lawn, golf courses, baseball fields, city parks, corporate front lawns, and anywhere else grass grows. They also grow in cemeteries, roadsides, pastures, and open fields. Meadow Mushrooms most frequently appear during the muggy days of summer when thunderstorms keep our lawns wet, but they can fruit into autumn if warm weather persists. If you live near farm pastures, it's a common find among those well-fertilized grasses.

COMMENTS: When Meadow Mushrooms appear in moist summer weather, conditions are right to encounter additional species of white-capped mushrooms in grassy areas—and those mushrooms might or might not be the Meadow Mushroom. Select each mushroom carefully, examining the gills and other features to make certain they match perfectly the description of *Agaricus campestris*. It's easy to get swept up in the excitement of collecting Meadow Mushrooms on a lawn and fail to notice that one of the mushrooms—or several—don't quite match the proper description.

In the Kitchen
with the Meadow Mushroom

Meadow Mushrooms, most chefs agree, have more flavor—a stronger one—than their store-bought cousin, the ubiquitous Button Mushroom. But there is a trade-off for this gift. Meadow Mushrooms can be a little more watery, and somewhat softer, in the skillet and not as meaty. But that's the only drawback. The flavor's sure worth it.

Prepare Meadow Mushrooms any way you would their store-bought relatives. Just don't serve them raw, as people often do (but shouldn't) in salads and on vegetable trays. Always cook wild mushrooms.

Meadow Mushrooms often show up in massive quantities, in every field, for miles, and we can't help but pick them all—or really try. So we bring home loads of mushrooms, more than we can eat or possibly share. But, like Halloween candy, we have no intention of sharing. And so we need to figure out how to horde our haul. Fortunately, Meadow Mushrooms dry nicely and can be used later to flavor sauces, soups, and gravies. They make mushroom sauce reductions so mushroomy, it's like mushroom espresso—but without the jitters or sleeplessness.

Try not to wash them before cooking; Meadow Mushrooms absorb too much water. Fortunately, Meadow Mushrooms aren't often seriously preyed upon by worms or other bugs. Because they were growing in grass, they should be soiled by nothing more than grass clippings. Just brush them off, inspect each one carefully, and then cut them up. That's it. Anyone who's ever cooked the basic Button Mushroom knows how to cook this one.

One reason this wild species of *Agaricus* isn't cultivated and sold commercially is because of its somewhat short life in the refrigerator. Use them up within a couple of days, or three days, tops—or dry them.

MEADOW MUSHROOM LOOK-ALIKES

Other Agaricus, Amanita thiersii, Chlorophyllum molybdites, Leucoagaricus naucinus

Agaricus sp.

Several species of *Agaricus* grow around trees, and while some are edible, some are known to be toxic. Therefore, avoid *Agaricus* found around trees, like this *Agaricus silvicola*.

Two poisonous species found in grassy areas look like drumsticks or golf balls on sticks when the mushrooms are young: *Chlorophyllum molybdites* (below) and *Amanita thiersii* (right).

Amanita thiersii.

Chlorophyllum molybdites.

Amanita thiersii.

Leucoagaricus naucinus is a potentially toxic, late-summer lawn mushroom bearing a superficial resemblance to the Meadow Mushroom. The differences: This species has white gills when young (compared with pink, then brown, gills on a Meadow Mushroom). Be careful to examine all of the various Meadow Mushroom traits for comparison because the gills of *L. naucinus* might also show a hint of pinkish gray as the mushroom matures—suggesting to the casual observer they've found a Meadow Mushroom.

The gracefully swollen stem base shown here is typical of *L. naucinus*, and a faint collar or ring (annulus) on the stem usually persists. Unlike the Meadow Mushroom, the smooth stem of this species bruises darker in variable shades of steel gray, sometimes with hints of blue or purplish-gray. In comparison, the stem of the Meadow Mushroom might bruise slightly translucent when crushed, perhaps very slightly yellowish, but no noticeable color change resembling the color gray should occur with a Meadow Mushroom.

Tip

How Important Is Gill Color?

Meadow Mushrooms have rosy pink gills when young, turning dark chocolate brown with age. Don't be fooled by any other gill color. The steel gray gills of this pretender should instantly prove you've picked a mushroom that definitely is not a Meadow Mushroom.

Various species of *Agaricus*, including a number of edible ones not described in this book, can have pinkish gray gills, or silvery gray gills, bordering toward white. But those are species different than *Agaricus campestris*—the Meadow Mushroom—which you can positively identify as edible because of its rosy pink, then chocolate, gill color. Some of the mushrooms with gray, silver, or white gills can kill you. What's more, the poisonous mistake might not be realized for many hours or a few days after the mushrooms were consumed. The onset of symptoms with deadly poisonous species is often delayed. Meanwhile, additional poisonous mushrooms might have been consumed if the individual decided the mushrooms were safe based on the lack of immediate symptoms (see "What Is Mushroom Poisoning?" p. 20).

Stropharia sp.

Brown Meadow Mushroom
Agaricus sp.

When people buy fresh mushrooms in stores, they often encounter three basic varieties: the common white Button Mushroom, a brown-capped "Crimini" (also spelled "Cremini") and the larger "Portobella" (sometimes spelled "Portabella" or "Portabello.")

Those trade names actually apply to the same species of mushroom: *Agaricus bisporus*. For years, the name *A. bisporus* meant only the white Button Mushroom. But then new strains of the same species were selected for their color and texture. Clever marketing people gave these strains new names—Portobella and Crimini—and suddenly a specialty industry was born.

In the wild, *Agaricus* species can be puzzling in their variations. The same species might look slightly different under different environmental conditions—or a slightly different one might actually be a new species. The Brown Meadow Mushroom is not the same species as the white-capped Meadow Mushroom. But if you appreciate the stronger flavor of Portobellas, you'll like this wild relative.

> **BROWN MEADOW MUSHROOM FEATURES**
>
> **CAP:** Medium-size, 2–4 inches across, light brown to tan with rusty, hairlike threads (but not scales) covering the entire upper surface, sometimes matted together to form darker tufts. Often darker near the center. Hairs might flatten and become less conspicuous with age, especially after rain, when the cap can appear light rusty brown overall. The cap can be somewhat blocky or squared when young, before it opens up to a convex, then nearly plane shape. When young, cottony threads connect the edge of the cap and the stem, forming a protective covering for the gills. As the mushroom grows, the cap expands and the protective veil tears apart, usually leaving a raggedy edge along the cap margin.
>
> **GILLS:** Narrow, rosy pink when young, becoming dark chocolate brown with age, nearly black. Gills stop short of being directly connected to the stem.
>
> **SPORE PRINT COLOR:** Dark chocolate brown (see "How to Make a Spore Print," p. 16).

STEM: Fairly short and stocky at times, arising from the grass barely high enough to allow the cap to expand. White to powdery silvery from the middle to the base, smoother, with a slightly darker, almost translucent sheen above that; rather soft (like the stem of store-bought Button Mushrooms), slightly fibrous. The stem bruises translucent, slightly yellowish, especially near the base. At the base of fresh stems, the interior flesh can appear fairly yellow when cut. In other species of *Agaricus*, that trait is usually a warning, since some woodland species known to be toxic also stain yellow at the base. Unlike the Meadow Mushroom, this species doesn't have much in the way of a ring on its stem, instead showing little more than a powdery or fuzzy ring.

HABITAT: All grassy areas, including your lawn, fields, pastures, and so on. Similar to Meadow Mushrooms, Brown Meadow Mushrooms most frequently appear during the muggy days of summer when thunderstorms keep our lawns wet, but they can fruit into autumn if warm weather persists.

On the Brown Meadow Mushroom, the ring on the stem will be faint and powdery, if visible at all; the base of stem might show some yellow color when bruised or cut; and the rosy pink gills turn dark chocolate brown with age.

COMMENTS: This is something of a new species to the literature in that it doesn't exactly match the scientific description of any known species of *Agaricus*. Its physical traits resemble a host of other dark-capped *Agaricus* species known across North America, including *Agaricus cupreobrunneus* and *Agaricus porphyrocephalus*. But, upon close examination, key traits of those species don't match this one. Scientists aren't fond of sticking a round peg into an octagonal hole, even if it kind of fits. Many people have eaten this "Pink Bottom" without any ill effects for years, yet science has yet to assign it a proper name. Pennsylvania *Agaricus* expert Richard Kerrigan generously examined specimens of this mushroom collected in southern Illinois in 2004 and matched its genetic code with a collection he made in Pennsylvania some years ago. The genus *Agaricus* is frustratingly

The Brown Meadow Mushroom has reddish brown hairs that often mat together to form scaly tufts. When young, the cap can have a blocky shape, opposed to the round shape of mature caps, and the cap usually doesn't rise very high above the grass because of the relatively short stem.

daunting in that everybody easily recognizes *Agaricus*, but almost nobody can confidently sort out all of the individual species. Science has yet to determine how many different species of *Agaricus* exist. This brown mushroom might someday prove to be nothing more than a regional strain of *Agaricus cupreobrunneus*, or it could be an entirely different species altogether.

Shaggy Mane
Coprinus comatus

One day they're a shaggy, egg-shaped cap on a stalk; the next, they're a ragged mess of ink. That's the quick life of *Coprinus comatus*, the largest of the aptly named "inky caps." Pick only fresh ones, before the dripping starts—and don't confuse them with other "inky caps." Even in the refrigerator, the Shaggy Mane will deliquesce into watery ink before tomorrow's supper, so cook up these distinctive treats soon after collecting them.

SHAGGY MANE FEATURES

CAP: Approximately the size and shape of a chicken egg when young, 2–3 inches tall, becoming columnar up to 6 inches tall; silver white with soft, scaly fibers and plates curling upward. Brownish, scale-filled center at top, often with bits of soil or gravel stuck to it as remnant debris from rising out of the ground. When young, the cap margin connects to the stem with a ringlike membrane or gasket that falls away as the cap curls upward with age. Before the cap can fully open, it begins to liquefy into a black ink, eventually dissolving almost completely, leaving just the stem and a raggedy cap relic.

GILLS: Thin and very closely crowded, silver white when young, becoming salmon pink before liquefying into black, spore-filled, inky nothingness.

SPORE PRINT COLOR: Black (see "How to Make a Spore Print," p. 16).

STEM: Hollow. Silver white, soft but fibrous, sometimes rising from the ground several inches. In large speci-

An inky drip from a liquefying Shaggy Mane.

TAKE THE FIELD

mens, the oblong-shaped cap might wobble on the stem like a loose lampshade as you hold it up to admire your find.

HABITAT: Favors grassy habitats, especially disturbed areas such as roadsides, but often wanders into gravel or seemingly desperate locations. It has been known to push itself up through marginal asphalt, causing scientists to study the amazing hydraulic pressure system employed by this soft, water-filled mushroom.

The edge of the cap of the Shaggy Mane dissolves into black ink as the mushroom ages.

The very crowded gills have a salmon pink edge above the inky parts.

The hollow, silky white stem splits easily.

COMMENTS: When young and fresh, the interior surface of the gills will be silvery gray. As the mushroom begins to liquefy, starting from the bottom, the lower margin of the cap turns salmon pink, then black, and finally into drippy ink. Look for the salmon pink color in the areas just above where the cap is becoming inky. Compared with other inky caps—and there are several other species out there, including toxic ones—this is the only one that shows the salmon pink color change. Also, look for the diagnostic, curly tufts on the shaggy cap. Those curls once led people to dub this mushroom the Lawyer's Wig, as in those white wigs worn by George Washington and still worn by members of British Parliament today.

Another inky cap, *Coprinus atramentarius*, commonly known as the Alcohol Inky Cap, lacks the salmon color change and also lacks the pronounced curls on the cap. The Alcohol Inky Cap contains coprine, which makes victims desperately ill if alcohol is consumed within a few days of eating the mushroom.

Because the wet, drippy ink of the Shaggy Mane actually makes a good writing ink, an intriguing story is sometimes told among mushroom storytellers: A certain agent of a certain government during a certain war—so the story goes—would send top-secret correspondence to officials using this very substance as ink. The receiver of the mushroom message, knowing the hidden access code, would examine the writing under a microscope to see if the very tiny, elliptical spores of *Coprinus comatus* were present in the ink. If not, the ink would be unmasked as regular, common ink and therefore branded a fake. They just don't make spies like they used to.

In rainy weather, the characteristic shaggy curls of the Shaggy Mane cap might appear flattened.

In less than twenty-four hours, a fresh Shaggy Mane can undergo a dramatic self-destruction process known as *deliquescence*. All but the center of the cap liquefies, leaving hungry mushroom hunters to long for yesterday.

SHAGGY MANE LOOK-ALIKES

Other inky caps, like this *Coprinus sp.*, have a brown, orange, or gray (not white) cap surface and silver white interior gill edges that remain that color and do not turn pink before they liquefy.

Clusters of several inky cap species, such as this Mica Cap (*Coprinus micaceus*), can be found throughout Illinois. And although some inky cap species are edible, some can be toxic. The Shaggy Mane (*Coprinus comatus*) is the only recommended, easy-to-identify inky cap in Illinois.

TAKE THE FIELD

Parasol Mushroom
Macrolepiota procera

If you manage to find a few Parasol Mushrooms in late summer or early autumn, look them over carefully to rule out mistaken identity with the notorious Green-Spored Lepiota (p. 21), which also can appear at the same time. There should be a swollen lump at the base of the stem, and the stem should be speckled with reddish scales. A raised brown bump should be noticeable at the center of the cap, as if the stem has been pushed upward, nearly poking through the cap. The spore print color should be white.

When you're satisfied all traits match the description in Features, your satisfaction will soon increase. Parasol Mushrooms are exceptionally delicious.

PARASOL MUSHROOM FEATURES

CAP: Size and round shape of the immature cap resembles a golf ball on a stick. Tan to light brown, but somewhat lighter—almost white—between the cracked scales, which break apart as the cap expands. In maturity, the cap becomes umbrellalike, then flattens (commonly 3–6 inches across), but always with a curved, slightly inrolled and scaly margin. A characteristic trait is the distinct, darker "nipple" centered on top. Overall surface is velvety smooth, like chamois or worn flannel, yet covered with soft brownish hairs or scales.

GILLS: White, becoming tan or very light brown. Thin, crowded, and often wavy or rippled. The gills are free (not attached) from the stem.

SPORE PRINT COLOR: White. Compare to the dingy, olive green spores of the toxic Green-Spored Lepiota (p. 21). Realize that deadly *Amanita* species also have white spores and free gills, so additional features must be examined. (See "How to Make a Spore Print," p. 16.)

STEM: Hollow or stuffed with cottonlike tissue, not solid; tending to be perfectly straight, like a pencil shaft, but sometimes with a mild curve on the lower portion. Very thin and graceful overall, tapering toward the cap. Stem is rusty tan on its exterior when very young, but soon develops rusty, reddish scales as the stem grows taller and expands, splitting and fracturing the reddish color into a mosaic of specks. A small lump or bulb should be visible at the base—but there is no volva or cup beneath the soil surface. Deadly *Amanita* mushrooms rise from an egglike sac underground, and one must gently dig below the base of the stem to find that important evidence.

The reddish brown scales and flecks on the very tall stem—plus the loose ring or collar—make the stem an important diagnostic part of the Parasol Mushroom.

The center of the cap on a Parasol Mushroom should have a dark bump.

TAKE THE FIELD

HABITAT: In grassy areas and forests, often near pines, but also hardwoods. That sounds contradictory, but it's where Parasol Mushrooms grow. Expect them in late summer and into fall, all fruiting simultaneously—when you find one, they're probably out everywhere, although they have a frustrating habit of growing as a solitary mushroom. They're not rare, but they're not as common as we'd like.

The Parasol's ring is often loose enough to slide up and down the stem.

COMMENTS: The velvety soft cap of the Parasol Mushroom rises above the ground with stately elegance, its scaly cap miraculously held aloft by a slender stem. When fully mature, the long stem really seems far too slender to confidently support the towering cap. It's like watching one of those circus acts where a man is balancing a coffee table on a pool cue—balanced on his chin. All of it defies our belief in the limits of balance. Confirming our suspicions, Parasol caps tend to topple over with age. If you're lucky, you'll catch a few before this quality act is lost. But there are other mushrooms whose stem can be tall and slender, including members of the genus *Xerula*, which have a distinct "root" underground, as well as some species of *Amanita*, which can be deadly. Parasol Mushrooms aren't difficult to identify once several traits are examined. Look for the brown, flattened scales on the cap, rusty speckles on the stem, a loose ring, free gills, white spores, and the lack of a volva. Parasol Mushrooms are truly delicious—but careless identification mistakes can be deadly.

Although often solitary, Parasol Mushrooms can be spotted from a distance because of the large cap with a dark bump in the center and the tall, slender stem. This one is from Mingo National Wildlife Refuge in Missouri.

With age, the Parasol's cap often topples over as the slender stem weakens.

The Parasol Mushroom has a swollen, bulblike lump at the base of the stem—but no cup or sac exists underground. This one is from Lake Glendale, Illinois.

The Poisonous Green-Spored Lepiota (p. 21) is often confused with the Parasol Mushroom. The loose ring resembles the ring on Parasol Mushrooms—be careful to examine all traits; making a spore print (p. 16) helps with identification.

TAKE THE FIELD

American Parasol
Lepiota americana

Rotting stumps and wood mulch are great habitats for several different species of *Lepiota*. The problem is, a few of those species are deadly poisonous. Not this one: the American Parasol can be identified by the combination of key traits—all of which must be absolutely obvious. Never eat this mushroom unless all traits match the description in Features. Eventually you'll realize that this delicious fungus—always associated with rotting wood—is actually quite easy to identify. But first you must learn to identify its collectively unique traits.

AMERICAN PARASOL FEATURES

CAP: Medium to large, 2–6 inches across. Tan or buff when young and still oblong or egg-shaped. As the cap matures and expands to a slightly conical saucer shape, the surface skin breaks apart into scaly, concentric bands, revealing the white flesh below. With age, the scales darken to a reddish rust color. A darker center, often raised in a bump (but usually less prominent than the "nipple" found on the Parasol Mushroom), remains throughout maturity. The cap turns somewhat darker—almost brick red—as it ages and dries.

GILLS: White, bruising first to a sickly yellow, then deep brick red (also with age). Gills are not attached to the stem.

SPORE PRINT COLOR: White (See "How to Make a Spore Print," p. 16).

STEM: Not as tall and slender as the Parasol Mushroom (p. 150), usually 3–5 inches tall, proportionate to the cap size. Slightly buff to tan with an overcast surface of faint, reddish scales. When young, the base of the stem should be obviously swollen, a helpful trait that persists with maturity. There should be no trace of a sac or volva beneath the stem underground, which is a trait of deadly *Amanita* mushrooms. When young, the edge of the cap and stem are connected by a white membrane that tears away from the cap as the mushroom matures and expands, leaving a ragged ring on the stem.

HABITAT: Most commonly found beside rotting stumps or in mulch, but also known to appear directly on logs which are in advanced stages of decay. It's one of the common urban mushrooms in Chicago. Mid-summer into the early fall are the most common seasons for finding one or two, or maybe fifty.

As the cap of the American Parasol Mushroom matures and expands, rusty reddish scales form rough, concentric rings around a darker center, which is raised in a slight bump.

COMMENTS: It's advisable to learn the traits of a deadly *Amanita* (p. 26) before picking and eating your first American Parasol. Essential traits of the American Parasol include the swollen, almost pear-shaped lower portion of the stem (which lacks an underground cuplike sac known as a volva), the color change when bruised, the typical habitat around rotting wood, and white spore color. Although some examples of the American Parasol don't grow very large—no more than 2 inches across the cap—the tendency is for caps to be 3–5 inches across, which is much larger than other, potentially toxic *Lepiota* species found in wood mulch habitat.

As the cap matures and expands, rusty reddish scales form rough concentric rings around a darker center, which is raised in a slight bump.

A veil-like membrane connects the stem and cap edge when the American Parasol is young, then remains as a collarlike ring as the cap expands, breaking the gasketlike seal.

An American Parasol's habitat is always near stumps or wood mulch.

Gills and stem bruise sickly yellow at first, eventually turning a deep rusty red. These American Parasols were refrigerated for a few hours after being handled. Bruised spots show red.

Young American Parasol Mushrooms arise straight from the ground or rotted wood. The swollen lower portion of the stem is especially noticeable when young. There should be no trace that the mushroom might have originated from an egglike structure.

AMERICAN PARASOL MUSHROOM LOOK-ALIKES
Other Lepiotas

Some species of mulch-inhabiting mushrooms related to *Lepiota americana* might resemble the American Parasol at first glance—and it's always helpful to examine other mushrooms for comparison. But the small size of the examples shown here should quickly give them away as something other than the larger American Parasol. Notice also the lack of a ring or collar on the stem of this impostor, as well as the fact the stem isn't swollen near the bottom. The closer we examine any mushroom, the more we learn how to safely identify edible species.

These small *Lepiota* mushrooms cannot be the American Parasol because of the reddish color present on the young, unbruised caps. Compare with the color of young, unbruised American Parasol caps on p. 156.

Also, some look-alikes have a slightly enlarged stem base (or a nuggetlike lump at the base). But, compared with potential look-alikes, only the American Parasol has an oblong, almost bowling-pin-shaped stem which should *not* be relatively thin, especially when young.

TAKE THE FIELD

9

Let's Eat

Recipes and Advice for Cooking Wild Mushrooms

Part of the joy of eating wild mushrooms is discovering the wildly different flavors and textures among the various species. Not all mushrooms taste alike. Some have absolutely no comparison whatsoever. It's like discovering there are actually other kinds of fruit besides tangerines, or other kinds of meat besides what's inside hot dogs. Prepare yourself for a big change as you sample the mushrooms in this book. If you think you know mushrooms based on your experience with marketplace exotics such as Portobella or Shiitake, you have so much more to discover.

In this chapter we present to you twenty-eight special wild mushroom recipes, many created for this book by top chefs from Illinois. Of course, not everybody wants to be the world's fanciest chef. But that doesn't mean everybody can't enjoy preparing these fantastic wild mushroom recipes. Some of them are fancy, some simple. Several recipes (A-1 Mushroom Sauce, Mushrooms over Toast, Wood-Stove Chicken Parmesan with Mushrooms, and Bacon and Eggs with Mushrooms) were designed with beginners in mind.

How Do You Cook Wild Mushrooms?

Any edible mushroom featured in this book can be tossed into a hot skillet with oil or butter, sautéed, then eaten with a sprinkle of salt. It's a great way to introduce yourself or others to a new mushroom flavor and texture before selecting an appropriate recipe. Not all mushrooms are compatible with all mushroom recipes.

For example, a meaty Chicken Mushroom (p. 38) can be sliced and cooked exactly like chicken breast. Thanks to its hearty texture, it's a near-perfect meat substitute. In comparison, the thin and delicate Black Trumpet (p. 109) isn't so versatile and is best used to flavor sauces or rice or egg dishes. Each mushroom has its own unique qualities and limitations in the kitchen, so the simple skillet test is recommended for beginners.

If there's any real trick to cooking wild mushrooms, it's knowing which flavors can overpower certain mushroom flavors. Smothering a Parasol Mushroom (p. 150)

with onions, for example, would overwhelm the velvet-smooth nuances of this choice edible. Frying the thin and chewy Wood Ear (p. 65) in beer batter would be a flavorless exercise in chewing. Experience and experimentation is the best teacher in the kitchen. You'll make mistakes along the way—but you'll learn.

About Mushroom Flavors:
Why Do Mushrooms Taste Different?

Carrots don't taste like onions, yet both are vegetables. A morel mushroom doesn't taste like a chanterelle mushroom, yet both are mushrooms. Plenty of mushrooms don't taste like other mushrooms, despite the fact they're all mushrooms. Sadly, many people don't realize there's a difference among all of those wild mushrooms out there—and sometimes it's a world of difference.

Just as beef doesn't taste like venison and grapes don't taste like peaches, the unique flavor and texture of individual mushroom species deserve consideration before selecting this or that recipe for a mushroom.

A surprising number of mushroom lovers—including some pretty good chefs—blink their eyes in confusion when informed that wild mushrooms are not a collection of interchangeable, complementary flavors. What's more, the same species of mushroom, collected in different locations, might produce very different flavors—better or worse. On that last point, some discussion is necessary.

A Yellow Chanterelle picked in northern Europe should not be compared to a Yellow Chanterelle picked in Illinois. For that matter, a morel picked in southern Illinois should not be compared to a morel picked in northern Illinois. There is a tremendous range of flavor quality and flavor intensity within the same species of mushrooms, depending on the region where the mushroom grows.

Why is this significant? Because the popularity of wild mushrooms is increasing, and with that popularity comes increasing demand for the "best" edible species. Unfortunately, certain mushrooms regarded as excellent in one part of the world might be only mediocre elsewhere—or lousy. Truffles, for example, are known to occur in the Midwest, but nobody digs them up because the truffles found here aren't worth eating. The fact is, mushrooms rated highly elsewhere might be startlingly unpalatable in Illinois, or vice versa. The authors favor *Craterellus cornucopioides* above all other Illinois wild mushrooms; yet that same mushroom is dismissed as potentially "bitter" in at least one European text. *Lactarius indigo* is sold in produce markets in South America, yet neither author can honestly recommend it beyond its culinary novelty as an edible blue mushroom.

With so many kitchen ingredients under scrutiny these days, an educated chef must now consider mushrooms based on geographical origin. Such judgments aren't new. Wine authorities long have realized that, when it comes to the world's important wines, what really matters is where the grapes are grown, not just the variety of grape. In the Bordeaux region of France, grapes produce legendary wines. The identical variety of grape grown in Chicago would produce, at best, fermented grape juice.

Why do the same species of mushrooms taste different? Simply put, taste differences are due to varying amounts of organic and inorganic ingredients. Not every mushroom produces the same amounts of the various chemicals that make it taste and smell like that species of mushroom. The longer answer is that mushrooms don't exist merely to be picked and eaten by humans. Their survival doesn't depend on it. The fact is, most species of mushrooms taste good to humans largely by coincidence. A fragrant metabolite produced by a fungus as a by-product of digesting wood isn't fragrant for the benefit of our human noses. Those great mushroom flavors we enjoy really might be just one of those lucky coincidences.

But why? Food experts demand answers. Alas, the true answer is complicated and fuzzy. Beyond the variation in climate and substrate, another critical factor that may influence the flavor of mushrooms is that mushrooms don't grow in a vacuum. Insects, slugs, small mammals, other fungi, and bacteria feed on mushrooms, and the mushrooms in turn produce chemicals to deter these voracious critters. For example, specific insects are attracted to specific mushrooms—and those insects might or might not be present in all regions where a particular mushroom occurs. Thus, the chemicals that attract or repel specific insects—chemicals that may be responsible for a scent or flavor humans might happen to like—might change from region to region as mushrooms modify their chemical composition in response to the local environment.

A fruit fly is attracted to the faint scent of apricots released by the mushroom *Cantharellus cibarius*. Mycophagous insects might help distribute spores long distances.

We quickly admit this is trivial to all but culinary nitpickers and bored entomologists. Really, there is no simple way to describe the complexity of factors that influence the flavor profile of mushrooms. Flavors and aromas are inconsistent and capricious properties within the still-mysterious mushroom, which continues to operate its chemical industries for reasons beyond the complete understanding of humans.

Although much remains to be understood about flavor variations in wild mushrooms, it's not essential to understand the reasons that, for example, Yellow Morel mushrooms taste better in northern Illinois than they do in southern Illinois.

But knowing there's a difference matters, especially for mycophagists—people who enjoy eating mushrooms. Understanding the range of flavors possible within various mushrooms makes for educated choices when preparing menus. The decision to use this or that mushroom for a particular dish should be based on known flavor and texture, not culinary fashion. The trouble is, we've become a society of culinary trivialities.

Some cooks use exotic ingredients merely to impress dinner guests, purchasing staggeringly priced rarities and presenting them in a dish—often with no discernible difference in flavor. A sprinkle of rare salt from an obscure Russian sea might dazzle our imagination, yet it's doubtful anyone could tell the difference between sodium chloride from Russia and sodium chloride from Utah when added to boiling pasta water.*

Of course, there's actually no crime in using uncommon ingredients for their own sake. It improves the culinary experience when we announce to guests before making the sign of the cross: "The precious tomatoes in this sauce came from my beloved Italian grandmother. The woman—God rest her soul someday—has been digging the same garden with her bare hands for eighty years. *Eighty* years."

Sure, interesting ingredients can be interesting for more than the flavors we taste in a great dish. But let's not forget the main ingredient of good food.

> * *We acknowledge all of you relentless dissenters here, the salt cultists, the ones now recalling excellent meals specifically involving a particular sea salt; please know that we support your extraordinary care in selecting ingredients. We really do.*

About Mushroom Powders

Potent Sprinkles That Improve Sauces, Soups, and Everything Else

Grind up dried mushrooms and you will get more than mushroom dust. What you've created is a secret kitchen ingredient known as a mushroom powder. Good chefs keep around a well-guarded jar of morel or maybe Black Trumpet powder to enhance the quality of special dishes. Powders can be made from almost any mushroom. Potent ones make the best powders.

Use a food dehydrator to dry the mushrooms bone-dry, then use a clean coffee grinder or food processor to pulverize them into dust. The result is concentrated mushroom flavor for any mushroom sauce or soup. Use it to pump up the richness of beef gravy or in mushroom risotto, or try it as a salt replacement on baked seafood. Sprinkle it anywhere a touch of mushroom flavor might help.

It takes a lot of dried mushrooms to produce a jar full of powder. But sometimes you'll find more mushrooms than you can eat fresh, and so you'll want to dehydrate some in your food dehydrator for this very purpose.

Don't forget to cook whatever you've sprinkled with mushroom powder. Drying mushrooms isn't an alternative to cooking them, and all wild mushrooms must be cooked.

LET'S EAT 163

Why Save One Mushroom?

Start Your Own Collection of Dried Mushrooms

Despite your best effort, you won't always fill your basket with edible mushrooms every time you go foraging for fungi. It's not your fault. Even the great mushroom hunters—mushroom hunters who can recite Latin names—come home once in a while with not a single mushroom to eat—or maybe one, just to save face.

But those great mushroom hunters make the best of their bad luck because they know the value of one mushroom. When it's dried and then added to an airtight jar where other orphans get stored, eventually there will be a multispecies stock of one-of-a-kind diversity. For special recipes, rehydrate a big handful and add them to a dish where you can savor the unique tangents of mushroomy flavors and textures—all in one.

Black Morel (*Morchella elata*).

But you must be truly positive of your identification. Figuring out which mushroom was misidentified becomes next to impossible once the meal disappears.

Semi-Dry Preservation

A Revolutionary Process to Keep Mushrooms Truly Fresh

Dried mushrooms can last years when stored properly in airtight glass jars. Mushroom hunters have been drying mushrooms since ancient times, and drying remains one of the best methods to preserve the complex yet fragile flavor of mushrooms. But there's a drawback: Dried mushrooms cannot be substituted for fresh mushrooms in all recipes. After being rehydrated, mushrooms tend to be chewier and therefore have reduced culinary benefits.

A simple process described here for the first time creates a partially dehydrated, frozen mushroom—stored in a vacuum—that retains the succulent properties of fresh mushrooms and allows us to enjoy "fresh" mushrooms for up to a year after they were picked.

This process was the result of much trial and error in the world's quest to freeze morels perfectly—and have them survive nearly as good as fresh.

"You're really on to something here," two expert chefs proclaimed independently after sampling wild morels that had been frozen eight months earlier. Both agreed the quality far exceeded that of any dried or frozen mushroom.

You'll appreciate the simplicity of this process in which mushrooms can be partially dehydrated, frozen in a vacuum, and later brought to the kitchen with much of the fresh quality intact. Here's how it works: Mushrooms are mostly water, which means the ice crystals that form during freezing inevitably rupture the majority of mushroom cells. That raggedy mess of tissue is why we get "mushy" mushrooms after thawing. By removing a fair amount of moisture from the mushroom *before* freezing, the remaining moisture in some of the tissue expands only slightly when freezing, rupturing a few cells here and there, softening the overall product but causing limited tissue damage. The softness in some ruptured cells compensates for those cells that dried completely and seem tough and chewy.

It's a balance between dried and frozen. The remarkable thing is, a chef can pull out a package of semi-dry mushrooms from the freezer and bring them right to the skillet. Use them any way fresh mushrooms would be used.

Here's a tip: Always freeze the mushrooms *before* sealing them in vacuum bags. Otherwise, partially dehydrated mushrooms, still-soft, crush in a vacuum. The frozen mushrooms, with just a trace of ice in the tissue, retain their shape in a vacuum—and store perfectly until next year's morel season. Don't forget to check the bags in the freezer once in a while, since those household-grade sealing machines sometimes don't seal perfectly.

Step 1: Clean and trim mushrooms and arrange on trays in a food dehydrator. Temperature settings (when available) should be no more than 110 degrees F.

Step 2: Dehydrate mushrooms until they become lighter in weight but not completely dry. Two or more hours is typical. A slight amount of moisture should remain. Place partially dehydrated mushrooms in the freezer for a few hours or overnight until remnant moisture freezes.

Step 3: Remove frozen mushrooms and quickly store in vacuum-sealed bags, and then return to the freezer. Note: There is no substitute for a vacuum-sealing machine. Ordinary freezer bags—even good ones—eventually allow air into the package, which leads to freezer burn and rapid deterioration of quality.

LET'S EAT

Chicken and Lobster Sunrise

Lasse Sorensen, executive chef
TOM'S PLACE, DESOTO, ILLINOIS

This luxurious breakfast requires something extraordinary: a perfectly young and tender Chicken Mushroom, not yet fully developed and firm. Unlike mature specimens, the tender meat of a velvety *Laetiporus cincinnatus* (the variety of Chicken Mushroom with a white, not yellow, underside) must be butter soft in order to blend successfully with the texture of eggs, creamy Asiago cheese, and lobster meat.

The mushroom itself isn't rare. Tough, old specimens survive outdoors for months. But anyone lucky enough to find the perfect Chicken Mushroom—which might happen only once every few years—will want to hurry to the lobster market and make preparations for this champagne-worthy breakfast.

Ingredients

LOBSTER AND EGGS

1 lb. whole lobster
3 Tbsp. Asiago cheese, shredded (or substitute Gruyère)
1 cup chopped Chicken Mushroom (p. 41)

1 palm-size lobe of fresh Chicken Mushroom
1 clove garlic, chopped
1 Tbsp. butter
2 oz. poached lobster claw meat, finely chopped
4 eggs
½ cup half-and-half (do not use heavy cream or whole milk)
Salt and pepper

SAUCE

1 cup cream
1 Tbsp. butter
Splash of lemon juice
1 tsp. chopped chives
Grated lemon rind zest

Preparation

LOBSTER AND EGGS

Steam whole lobster until cooked, about 15–20 minutes, remove claws and claw meat. Keep remaining lobster whole and warm. In a medium skillet (nonstick omelet pans are best), melt butter, sauté garlic and chopped Chicken Mushroom for 1 minute, and then add lobster claw meat. Whisk together eggs and half-and-half, add salt and pepper to taste, and add to skillet. When eggs begin to firm, fold and cut into two portions.

SAUCE

In a skillet over medium heat, reduce cream, cheese, butter, and lemon juice until thickened. Add grated lemon rind zest, salt, and pepper to taste.

Serve as illustrated by laying omelet portion on one side of the plate, the lobster tail on the other, and the luscious cream sauce everywhere you want it. Serves two.

Wild Turkey and Morels

Lee Conway, executive chef
CONWAY'S CATERING, BELLEVILLE, ILLINOIS

The successful wild turkey hunter walks slowly from the spring woods with one additional hope—a big one—that there will be at least fifty morel mushrooms between here and there, approximately. The route between here and there, of course, will be connected by backward loops and zigzags, which is pathetic—but understandable. A wild turkey hunter must have fresh morels to celebrate a wild turkey properly, and this dish is the one. Brandy and cream make the sauce; the breast is pan-fried.

Ingredients

TURKEY

½ wild turkey breast, cleaned and skinned
1 cup flour
1 tsp. salt
1 tsp. pepper
2 Tbsp. butter

MORELS

2 Tbsp. butter
1½ cups fresh morels, cleaned, with most of stem removed
1 tsp. garlic, minced

SAUCE

½ cup chicken stock
½ cup heavy cream or half-and-half
3 Tbsp. butter
3 Tbsp. flour
3 Tbsp. brandy

Preparation

TURKEY

Slice turkey breast across the grain into ¼-inch medallions, then dust each medallion with a combined mixture of the flour, salt, and pepper. Melt butter over medium-high heat, add turkey medallions, and brown on each side, which should take no more than 1 minute for each side. Set aside and keep very warm on a hot plate.

MORELS

Split lengthwise. For large morels, cut into thick rings or bite-size pieces. Over medium-high heat, melt butter and then add morels and garlic, cooking briskly for several minutes until morels release their liquid and begin to pop. Set aside and keep warm.

SAUCE

Combine chicken stock and cream or half-and-half and warm over low heat. In a separate skillet, melt butter over low heat. Slowly add flour, mixing into a thickened base known as a roux. Carefully and slowly whisk the warm stock and cream into the pastelike roux, simmering over low heat until thickened. Add brandy and the cooked morels, and ladle over the turkey medallions.

Serve immediately. Serves two.

Cooking Morels
Tips for Making the Most of the Prized Sponge-Cap

Executive Chef Lasse Sorensen cooking morels at Tom's Place in Desoto, Illinois.

Since morels are the most popular wild mushroom in Illinois—yet usually rare and precious—everybody asks: "What's the best way to cook morels?"

Nobody wants to ruin a perfectly good batch of these spring delicacies. Such disasters won't happen if you understand some morel basics: Morel stems can be a little tough and have much less flavor than the caps, so trim away most of the stem (but save them for stock or dehydrate them to concentrate the flavor). The caps can be sautéed in butter, with a touch of garlic, and can go straight into virtually any mushroom recipe. Cook morels well. Volatile compounds in morels need to be cooked away to reduce the risk of anyone getting sick.

Morels often contain forest debris, including insects, within the pitted chambers of the cap and within the hollow interior. Brush away what you can, and dunk them in cold water (no salt) if necessary. Many chefs insist mushrooms must never be washed, since mushrooms already contain mostly water. But as long as morels are allowed to dry for an hour or more before cooking, they're as good as any morel you might pick the day after it rained.

Double Oyster Chowder
Lasse Sorensen, executive chef
TOM'S PLACE, DESOTO, ILLINOIS

Oyster Mushrooms and real oysters actually have a thing or two in common—not identical. But if oyster chowder is your thing, why not toss in a load of freshly sautéed Oyster Mushrooms to keep the oyster party going? Oyster Mushrooms got their name, we're pretty sure, because the caps are basically oyster shaped. It's not the taste. But it's also not a stretch to suggest common tendencies: Both can be a little chewy, and both flavors remind us of similar food experiences. It's a natural match—and here's a bonus: For those of us who are just so-so with real oysters, those bites of delicious Oyster Mushroom keep everybody happy to the bottom of the bowl.

Ingredients

- 5 Tbsp. butter
- 1 tsp. garlic, chopped
- 3 cups fresh Oyster Mushrooms, chopped (or substitute dried) (p. 46)
- 1 cup dry white wine
- 1 quart chicken stock
- 1 cup water
- ½ quart cream
- flour
- Salt and pepper
- 3 oysters per person
- Chopped parsley and diced tomato bits for garnishes

Preparation

In a large skillet, melt 2 Tbsp. butter over medium-high heat and add chopped garlic and mushrooms, stirring until juices are released and mushrooms are lightly browned. Add wine, simmer until liquid volume is reduced by one-third, and add chicken stock, water, and cream. To thicken, melt 3 Tbsp. butter in a skillet over medium heat, stir in enough flour to make a thickened base known as a roux, and then slowly whisk the roux into the chowder. Season with salt and pepper to taste. Add fresh oyster meat and simmer for 2 minutes—no more—until oysters are just firm. Tough and chewy oysters will result from lengthy cooking. Ladle into soup bowls, garnish with chopped parsley and diced tomato bits, and serve.

Note: Dried Oyster Mushrooms can be substituted for fresh by first reconstituting the dried mushrooms in enough chicken stock to cover the mushrooms. After 5 minutes, remove mushrooms, squeezing out excess liquid, then sauté as directed for fresh mushrooms.

Nobody follows recipes exactly when making chowder. Add scallops or shrimp if you prefer. Go heavy on the mushrooms if you've hit a bonanza from the wild, or be stingy with mushrooms if you're reduced to buying those expensive cultivated versions at the market.

Black Trumpet Salad

Joe McFarland
MAKANDA, ILLINOIS

Almost everybody's tried fresh spinach salad with hot bacon dressing—very good.

Now, imagine a cold salad plate, a chilled fork, fresh-picked garden lettuce . . . then, like a kitchen sucker punch, *warm* Black Trumpet mushrooms—with cool blue cheese dressing. Hot and cold. Strong with mild.

Try it and you'll agree this should be on every major restaurant menu in the world. The essential point for success is keeping the plate and fork very cold until the moment the Black Trumpets are cooked. Rush the hot mushrooms to the chilled salads and serve immediately, without hesitation or delays. There should be no temptation to preclean in the kitchen, no rinsing of pans or quick counter-wiping. Sit down and eat. Avoid fumbling with corkscrews, wasteful speeches, or extra-napkin fetching—you must eat this salad now. Telephones should be disconnected in advance, and all door handles propped against entry with heavy chairs.

Ingredients

FOR EACH SALAD

Fresh garden lettuce (use mild varieties; avoid stronger-flavored greens)
1 cup fresh Black Trumpet mushrooms (p. 109)
2 Tbsp. butter
¼ cup white onion, sliced
3 Tbsp. cold blue cheese dressing (add crumbled blue cheese, if desired)

Preparation

Chill salad plates and forks for at least a half-hour; very cold is best. Wash and drain lettuce and arrange on chilled plates and return to refrigerator. Over medium-high heat in a skillet, quickly sauté Black Trumpets in butter or oil for 3–5 minutes, until almost limp, then add onion slices and sauté an additional minute until onions begin to turn translucent. Very quickly scatter hot Black Trumpet mixture on cold salads, adding chilled blue cheese dressing. Serve instantly. A crisp Riesling matches nicely.

KITCHEN TIP: Keeping the forks and salad plates very cold—not merely cool—until the crucial last second is absolutely essential. Otherwise, the warm lettuce and mushrooms are about as appealing as a warm glass of water.

• •

Steak Mushrooms

Mark Fontana, executive chef
BOGEY'S AT STONE CREEK, MAKANDA, ILLINOIS

If there's anything better than a good slab of beef under a sizzling heap of freshly sautéed mushrooms, we really don't know it. Meat lovers see this combination on a plate and begin growling, unleashed, like a werewolf pawing at his necktie. The animal power of mushrooms and steak is what makes us suddenly believe we can eat an eighty-ounce prime rib in one of those joints where anyone man enough to get the job done wins it all for free. Almost any mushroom works with this Worcestershire-influenced steak topping. The Purple-Gilled Laccaria (p. 73) is a strong favorite. Think venison, lamb, or beef—any red meat will do. Now, move out of the way, sissy.

Ingredients

1½ quarts beef consommé
1 clove garlic, sliced thin
1 bay leaf
½ cup water
¼ cup Worcestershire sauce
1 cup tomato juice
2 Tbsp. butter
1 lb. assorted wild mushrooms, roughly chopped
½ Tbsp. cornstarch, mixed with a little water
Salt and pepper

Bogey's prime rib with *Laccaria ochropurpurea*.

Preparation

In a large pot, combine all ingredients except butter, mushrooms, cornstarch mixture, and salt and pepper. Bring to a slow simmer for 15 minutes, allowing flavors to blend. In a large skillet over medium-high heat, melt butter and then add mushrooms. Sauté 5–7 minutes or until browned, seasoning to taste with salt and pepper. Bring consommé mixture to a boil and thicken slightly by stirring in cornstarch mixture. Ladle the hot consommé into the pan of mushrooms and cook until reduced slightly. Adjust seasoning, then serve over prime rib, steaks, chops, etc. Serves ten.

. . . .

Black and Bleu Morels

Mark Fontana, executive chef
BOGEY'S AT STONE CREEK, MAKANDA, ILLINOIS

Fans of pungent, rich cheeses and strong, powerful flavors will love this fungus-enriched steak partner. A good beef or venison steak, covered with Black Morels and a hearty pour of this rich blue cheese sauce, will—and this is no joke whatsoever—instantly improve your opinion of being knocked out.

Everybody thinks blue cheese is impossible to match with big flavors. "Doesn't blue cheese overpower the morels?" everybody asks, logically. They've never tried this combination. Here's the surprise: Blue cheese and Black Morels—the strongest-tasting of the Illinois morels—complement each other supremely well. Dinner guests will gasp, chewing will stop—and eyes will search for witness, it's that good.

Black Morels are a bit too strong for some people; ditto for blue cheese. If you're not in this hard-core league, bypass this recipe.

Ingredients

2 Tbsp. salted butter
1 or 2 cloves garlic, crushed
2 cups Black Morels (p. 89)
1 cup heavy cream
¾ cup blue cheese, crumbled

Preparation

This robust blue cheese sauce can be prepared with fresh or dried Black Morels, with just a slight modification of steps. When using dried morels, soak them first for 10 minutes in just enough warm water to cover, then drain and pat the morels dry. If using fresh morels, split them lengthwise or keep them whole if you're confident of a clean interior.

In a large skillet, melt butter over a medium-high heat and add garlic and mushrooms. Cook for several minutes until morels are browned and begin to pop. Remove pan from heat but keep warm. In a medium saucepan, simmer cream until bubbly and reduced to the point where a spoon cutting across the bottom of the pan leaves a clean line, if only for a moment or two. Add the blue cheese crumbles, simmering for a very short while, keeping the blue cheese chunky and not reduced to a smooth consistency (a chunky blue cheese sauce is always better than a smooth blue cheese sauce, for hard-to-explain reasons—basically the same way that two slices of bacon always taste better than one slice of bacon). Combine the morels and the sauce and serve over prime rib or steak.

COMMENTS: The stems of morels are less flavorful than the caps, so using mostly the cap portion will impart the best flavor. Do not be tempted to try this recipe with Yellow Morels. The strong flavor of blue cheese will smother the great flavor of Yellow Morels, whereas the flavor of Black Morels matches perfectly.

Beer Batter Morels with Wild Garlic Mustard Dipping Sauce
Andrew McGovern, executive chef
BACKSTREET STEAK HOUSE, GALENA, ILLINOIS

Why bother to improve on something as perfect as deep-fried morels? Everybody loves fried morels—especially with ranch dressing for dipping—and the absolute simplicity is part of the beauty. It's a classic. But here's a new twist: Fry your morels in this basic beer batter, then mix up this mayonnaise- or sour cream–based dipping sauce made from edible, wild garlic mustard leaves.

Garlic mustard, for those unfamiliar with it, is an exotic plant now found in nearly every forest in Illinois. It was introduced to North America more than 150 years ago and, for decades, was rarely noticed. Inexplicably, garlic mustard populations exploded during the 1990s, smothering wildflowers and rare plants, becoming a major ecological threat in our forests. Naturalists now hate it. But those garlicky, diced-up leaves sure make a fine-flavored sauce for fried morels.

Ingredients

MORELS

12–36 fresh morel mushrooms, cleaned
1 quart cooking oil for deep frying

BEER BATTER

1 bottle of good beer, 12 oz.
1½ cups flour
1 tsp. baking soda

GARLIC MUSTARD DIPPING SAUCE

1 cup mayonnaise or sour cream
⅛ cup chopped fresh garlic mustard leaves
Squeeze lemon juice

Preparation

Make the dipping sauce first by mixing all ingredients and chilling in a serving dish. While cooking oil is heating, mix beer batter ingredients together and clean morels so they're ready for frying. When oil temperature reaches 365 degrees, dip morels into batter and then drop into hot oil. Depending on the size of the deep-frying skillet or kettle being used, adding too many mushrooms at once will drop the oil temperature below an acceptable frying temperature, resulting in soggy mushrooms. When mushrooms are dark golden brown all over, remove from oil and drain. Serve immediately with dipping sauce and a bottle of good beer.

FRYING TIP: Ever notice how fast-food employees lift up a basket of cooked French fries, then hang the basket above the hot oil for a minute before dumping the fries into a wide bin for salting? Hanging a fry basket over hot oil not only recovers a few drops of oil, but the rising heat seems to draw out excess oil from whatever was being fried as well, resulting in crispier fries—or mushrooms.

. . . .

Scallop-Hericium Checkerboard

Sean Keeley, executive chef
INDIGO, SPRINGFIELD, ILLINOIS

The lightly sweet flavor and texture of the white autumn mushroom known as Lion's Mane (*Hericium erinaceus*) reminds some people of scallops. The seemingly unlikely pair are brought together here in this eye-catching presentation with grilled, basil-stuffed Roma tomatoes and a balsamic syrup.

This delightfully inventive appetizer is further evidence that all mushrooms are not created equal. Your dinner guests will have trouble figuring out how a mushroom can match the color, flavor, and texture of a scallop, leading to the inevitable chicken-versus-the-egg dinner conversation: "Does a scallop taste like *Hericium*—or does *Hericium* taste like a scallop?"

Ingredients

SAUCE

2 cups balsamic vinegar
1 Tbsp. honey

SCALLOPS

6 large dry scallops (U10)*
3 Roma tomatoes
6 large basil leaves, plus a few extra sprigs for garnish
6 pieces Lion's Mane mushroom, cleaned (p. 59)
1 Tbsp. olive oil
Salt and pepper

*U10 is a size count meaning fewer than 10 per pound. Dry scallops, or diver scallops, are packed fresh from the shell with no water or phosphates added. Use these if available; if not, frozen will do.

Preparation

SAUCE

Simmer the balsamic vinegar in a 2-quart pot over medium heat until reduced to the consistency of maple syrup. Add honey and store at room temperature for up to 8 months. This is a great all-purpose sauce for steaks, pasta, or anything else—even fresh strawberries. Warning: A little goes a long way, and the smell while cooking can take your breath away. But it's well worth the effort. Keep an eye on the pot and reduce heat if vinegar starts to boil rapidly. Scorched balsamic vinegar is no fun to smell or clean.

SCALLOPS

Pat dry the scallops. Slice the ends from the Roma tomatoes, then cut in half. Remove the seeds from the tomatoes and stuff the cavities with torn fresh basil leaves. Cut the cleaned mushroom into scallop-size chunks. Heat olive oil in a 12-inch skillet over medium-high heat. Season scallops, mushrooms, and tomato with salt and pepper and then add to skillet slowly—scallops first, then mushrooms, and finally the stuffed tomatoes. When scallops are well-browned on one side (about 2½ minutes), turn everything over and cook another 2–3 minutes. Chef Keeley prefers scallops a bit rare, and removes the pan from the heat after turning over the scallops, letting the items rest in the skillet for a few minutes.

ASSEMBLY

Drizzle two plates with a little of the balsamic syrup. Arrange scallops, mushrooms, and tomatoes in a checkerboard design, garnish with a fresh basil sprig and serve immediately. Serves two.

. . . .

Chicken Laccaria
Justin Hagler, executive chef
MARY'S, HERRIN, ILLINOIS

A working-class mushroom like *Laccaria ochropurpurea* has a strong texture but very mild flavor. Experienced mushroom hunters know this, and so sometimes ignore them. Therefore, surprise your know-it-all mushroom friends with this elegant redemption that transforms the ordinary *Laccaria* into culinary greatness. It's a blend of grilled, chicken-wrapped mushrooms and goat cheese under a brandy-thyme cream sauce. Technically simple, it can be prepared in 45 minutes. Nobody needs to know the richly delicious mushrooms on their fork don't always taste this incredibly good.

Ingredients

CHICKEN

8 Purple-Gilled Laccaria mushrooms, finely diced (p. 73)
8 Purple-Gilled Laccaria mushrooms, whole
4 boneless, skinless chicken breasts, 4 oz. each
2 cups goat cheese
12 asparagus spears
Vegetable oil

SAUCE

3 Tbsp. butter
2 shallots, minced
8 Purple-Gilled Laccaria mushrooms, sliced
3 sprigs fresh thyme
3 oz. brandy
Pinch white pepper (or to taste)
Pinch cayenne pepper (or to taste)
Pinch nutmeg (or to taste)
1 pint heavy cream
Salt

Preparation

CHICKEN

Finely dice 8 mushrooms for stuffing and save 8 for roasting. Using a tenderizing mallet, pound chicken breasts to ¼-inch thickness. Place flattened breasts, smooth side down, on a surface for stuffing. In a bowl, combine goat cheese and diced mushrooms, dividing into four sections. Stuff chicken with goat cheese mixture and roll up tightly. Grill until marked. Brush stuffed chicken, asparagus spears, and whole mushrooms with oil and place in a 350 degree oven for approximately 20 minutes or until internal temperature of chicken reaches 165 degrees.

SAUCE

Over medium-high heat, melt butter and cook shallots until opaque. Add sliced mushrooms, thyme, brandy, white pepper, cayenne pepper, and nutmeg. Flame brandy to burn off alcohol. Next, add cream. Reduce heat and simmer until thickened. Add salt and adjust spices, if necessary. Serve one stuffed chicken breast over three asparagus spears, ladle a healthy portion of sauce, and top with roasted mushroom caps. Serves four.

A-1 Mushroom Sauce

Loyce Bleichner

PIZZA SHACK, BENTON, ILLINOIS

This simple mushroom sauce is especially welcome over a cut of meat that is less than perfect or a bit tough. The sauce really improves the experience.

Ingredients

1 lb. fresh sliced wild mushrooms (just about any kind will do)
1 thinly sliced onion (Vidalia-type sweet onions are best)
1 clove garlic, minced
½ cup butter
1 cup A-1 steak sauce

Preparation

Sauté mushrooms, onion, and garlic in butter until mushrooms are well-browned. Add A-1 sauce and simmer until slightly thickened. Serve over your favorite steak.

Mushrooms over Toast

A simple, traditional recipe.

Ingredients

3 Tbsp. butter
2 cups fresh wild mushrooms, cut into bite-size chunks
Salt and pepper
2 Tbsp. flour
2–3 slices toast

Preparation

In a medium skillet, melt 2 Tbsp. butter over medium-high heat. Add mushrooms and a pinch of salt and pepper to taste. Cook, stirring occasionally, until mushrooms are well-browned. Add remaining Tbsp. butter and sprinkle flour over mushrooms, stirring until mixed well and thickened. Serve immediately over hot toast on warm plates. Serves two to three.

Mushrooms with Bacon

Another simple, traditional recipe

Ingredients

½ lb. bacon
2 cups fresh wild mushrooms
2 Tbsp. flour
4 slices toast
Salt and pepper

LET'S EAT

Preparation

In a large skillet, fry bacon over medium heat until bacon is nearly crispy. Remove bacon but do not drain skillet. Add mushrooms and salt and pepper to taste, cooking over medium-high heat. When mushrooms are golden brown, mix with bacon and flour, stirring until thickened, and serve immediately over hot toast as a breakfast side dish. Serves four.

Wood-Stove Chicken Parmesan with Mushrooms

This energy-efficient main course is a backwoods version of basic Chicken Parmesan. But even if you cook this easy-to-make meal on a modern stove, the flavorful memory of summer mushrooms is worth the utility bill.

Ingredients

2 frozen, skinless chicken breasts
Salt
1 cup pasta sauce (store-bought saves time)
1–2 handfuls dried mushrooms (Meadow Mushrooms, boletes, or Black Trumpets)
2 Tbsp. butter or oil
¼ tsp. garlic powder
Parmesan cheese

Preparation

CHICKEN AND SAUCE

Place frozen chicken breasts in a large cast-iron skillet, sprinkle with salt, cover, and place over low heat (the top of a wood-burning stove is ideal). Chicken breasts should thaw and begin to simmer in their own juices within 10 minutes. Depending on the stove heat, chicken should take up to an hour to cook thoroughly. Meanwhile, heat pasta sauce and keep warm.

MUSHROOMS

In a glass bowl or crock, add dried mushrooms and enough water to cover. Place near edge of wood stove to allow mushrooms to soften in the warm water. Drain mushrooms and gently squeeze out excess moisture (save the liquid for some other project if you like). In a medium skillet, melt butter or oil over medium-high heat (a regular stove works best here). Add mushrooms and garlic powder and sauté until golden brown. To serve, place cooked chicken breast on a plate, cover with pasta sauce, top with mushrooms, and sprinkle Parmesan cheese over everything. Serves two.

Bacon and Eggs with Mushrooms

Ingredients

6 strips bacon
1½–2 cups fresh wild mushrooms, sliced
5 eggs
¼ cup whole milk
Salt and pepper
2 slices toast

Preparation

In a large skillet, fry bacon until crispy or soft (your preference). Remove bacon, crumble, and drain most of the drippings. Add mushrooms to skillet and cook until juices are released and lightly golden. Combine egg, milk, and salt and pepper to taste, add to mushrooms along with bacon crumbles, and scramble over medium-high heat. Serve over hot toast. Serves two.

Puffball S'mores

Joe McFarland
MAKANDA, ILLINOIS

Cutting into a fresh puffball is strangely like cutting into a giant marshmallow—but without the stickiness. If a marshmallow could be a mushroom, it would be a puffball, which brings us to this impossible leap: Puffball S'mores.

What better way to salute the campfire classic than to switch the mushroom for marshmallow, meat for chocolate, and bread for Graham crackers? A layer of melted Swiss cheese covers this bacon-and-mushroom party pleaser. The quantities can all be adjusted according to need.

LET'S EAT

Ingredients

Bacon
Giant Puffball mushroom sliced ⅓-inch thick (p. 130)
White bread
2 Tbsp. butter
Salt
Swiss cheese
Chives

Preparation

Cook bacon until done but not crisp and drain. Using a round cookie cutter, create equal quantities of puffball and bread disks. Melt butter in a skillet over medium-high heat and sauté mushroom disks until mushrooms are lightly golden. Add salt to taste. Remove from skillet and set aside. On an ungreased baking sheet, lay out as many bread disks as desired and stack layers of cooked puffball, cooked bacon, and Swiss cheese on top of the bread disks. Bake in 400 degree oven for 5–7 minutes or until cheese melts. Serve immediately with cold beer and football.

Cool Oyster Salad with Ginger Sauce
Bao Cheng Lee, executive chef
SAO ASIAN BISTRO, MARION, ILLINOIS

Steamed Oyster Mushrooms chilled and served with a zesty ginger sauce make this cool salad a lively accompaniment to Asian cuisine. Flecks of red pepper spice it up on the cold plate. Serve it with a sprig of fresh cilantro and toast yourself with hot tea. You've never had Oyster Mushrooms like this before.

Ingredients

4 cups fresh, firm Oyster Mushrooms (do not use dried)
¼ cup dry sherry
¼ cup V-8 vegetable juice
¼ cup soy sauce
¼ cup vegetable oil
2 tsp. sugar
1 clove garlic, grated
⅛ cup ginger root, grated
½ onion, diced
2 Tbsp. chili oil
2 Tbsp. apple cider vinegar (or other sweet vinegar)
Sprigs of cilantro for garnish

Preparation

Steam mushrooms with salted water for about 5 minutes (or until cooked), rinse in cold water, and chill. Combine all remaining ingredients for the ginger sauce. Arrange chilled Oyster Mushrooms on a cold plate, cover with ginger sauce, garnish with sprigs of cilantro, and serve immediately. Serves four.

Enchanted Grifola

Bao Cheng Lee, executive chef

SAO ASIAN BISTRO, MARION, ILLINOIS

This centerpiece for an autumn banquet relies upon that greatest of discoveries: A perfect Hen-of-the-Woods, fresh and clean and without any sign of serious intrusions from alien life. This steamed bouquet of *Grifola frondosa*—also known as Maitake—draped in a soy-based garlic brown sauce, will honor any occasion you can invent. Maitake is known as the dancing mushroom, perhaps for this very reason. Carve slices of it onto beds of fried rice for your guests, and modestly acknowledge the due credit for finding the perfect fall mushroom.

Grifola frondosa appears in the autumn under oaks, and, if your timing is right, you might discover a perfect one, without a blemish on it anywhere. Unfortunately, years might pass before you find that perfect one. A full cluster of Hen-of-the-Woods can be utilized for this recipe (smaller, young specimens are best), providing the mushroom is free from organic debris, insects, and so on. Because perfectly clean and fresh specimens of *Grifola frondosa* are rather uncommon, soaking the mushroom in lightly salted water for 15–20 minutes can help evict unwelcome guests from within. As an alternative, a portion of a cluster can be cut out and utilized, making it possible to visually inspect the interior. If salted water is used to clean the mushroom, skip adding salt to the water for steaming.

Ingredients

Hen-of-the-Woods mushroom, cleaned and/or soaked (p. 35)
¼ cup soy sauce
1 Tbsp. diced garlic
1 Tbsp. cornstarch
1 Tbsp. sugar
1 Tbsp. sesame oil
½ cup white wine

Preparation

Steam the cleaned mushroom in lightly salted water for 7–10 minutes or until cooked and tender (use unsalted water if mushroom has been previously soaked in salted water). Drain. For the garlic brown sauce, combine the remaining ingredients and simmer until thickened slightly. To serve, arrange mushroom on a platter with your choice of garnish and cover with a generous helping of the sauce. Individual portions can be served over stir-fried or plain white rice.

Chile Oyster Soup

Bao Cheng Lee, executive chef

SAO ASIAN BISTRO, MARION, ILLINOIS

It is often said that homemade soup is better on the second day, after the flavors blend together—but not here. This original soup medley demands freshness when it comes to individual flavors that work flawlessly well together: Oyster Mushrooms, fresh tomato, basil, and a red-pepper chaser. Put them together in a light chicken broth and serve immediately for this easy-to-spoon Asian import.

Ingredients

    4 cups chicken stock
    1 cup shrimp, cleaned and deveined
    1 lemongrass
    Juice from 1 or 2 limes
    2 Tbsp. fish sauce
    3 chile peppers, crushed
    3 kaffir lime leaves
    2 cups fresh Oyster Mushrooms, sliced (p. 46)
    5 sprigs cilantro
    1 Tbsp. nam prig pow
    Tomato wedges, basil leaves, and lemon slices for garnishes

Preparation

Freshness is a key element to success, so you'll want to have the serving bowls ready. In a stockpot, bring chicken stock to a boil as you clean the shrimp. Cut lemongrass into 6-inch pieces and bruise or crush lightly to release the flavor. Add the lemongrass to the stockpot and boil for several minutes. In each serving bowl, add equal portions of the lime juice and fish sauce. Add crushed chile pepper to each bowl. Into the stockpot add the kaffir lime leaves (all stem parts removed), plus the Oyster Mushrooms. Add shrimp, then remove from heat almost immediately, before shrimp get tough. Shrimp can either be removed from the soup after releasing their flavor or retained. Add soup to individual bowls, add nam prig pow and garnish with fresh tomato wedges, basil, and lemon. Serves two to three.

. . . .

Mushroom Bolognese

Mark Fontana, executive chef
BOGEY'S AT STONE CREEK, MAKANDA, ILLINOIS

Everyone should know how to make a great Italian meat sauce. They're an indispensable base for so many Italian dishes—like that *really* good pasta sauce you had at a dinner party once, or that perfect lasagna from that one restaurant. Those chefs knew how to make a great Bolognese, the fundamental meat sauce.

A reduction of white wine and milk is the secret for this buttery-smooth Bolognese. Add ground beef or venison to start and sautéed wild mushrooms to finish. Feel free to use almost any edible wild mushroom you find. Unless you're in college, you'll never open a jar of "pasta sauce" again.

Ingredients

4 oz. diced pancetta (salt pork or bacon will substitute)
1 cup olive oil
1 lb. onion, diced
8 oz. celery, diced
Salt and pepper
4 lbs. ground beef (or a mixture of ground beef, venison, and/or pork)
3 cups white wine
3 cups whole milk
Pinch fresh nutmeg
3 lbs. assorted wild mushrooms, chopped
5 lbs. canned plum tomatoes, drained

Preparation

In a large skillet, render pancetta over low to medium heat for 6–8 minutes and remove from pan. Add ½ cup olive oil, onion, and celery. Cook over medium to medium-high heat for 12 minutes. Do not brown (cook just until onions are translucent). Add salt and pepper to taste. Add ground meat mixture, breaking apart as it cooks, mashing if necessary to obtain a fine texture—the finer the better. Cook 6–8 minutes until meat is no longer pink, but do not brown. Move meat to one side of pan and tilt pan slightly to remove as much of the fat as possible. Adjust seasoning with salt and pepper if desired. Add wine and reduce heat to a bare simmer and cook until almost dry, about 15 minutes. When wine has evaporated, add whole milk and nutmeg and simmer lightly until almost dry. Remove pan from heat. Heat another large skillet with remaining 1/2 cup olive oil, add mushrooms, and cook until golden brown. Season with salt and pepper (you might want to cook mushrooms in batches). Drain tomatoes, reserving juice (use juice to thin out sauce, if desired), and crush tomatoes by hand. Add tomatoes and mushrooms to the meat sauce and cook slowly until meat is soft and sauce has thickened slightly. Serve over any pasta or use the meat sauce for baked pastas. Serves twelve.

Tempura Morels

Andrew McGovern, executive chef
BACKSTREET STEAK HOUSE, GALENA, ILLINOIS

We don't often suggest soaking mushrooms, but soaking morels in lightly salted water before fixing a tempura batter for deep-frying will make the flavor of the morels really pop—especially when swabbed with a dipping sauce of wasabi cream and a balsamic-port reduction. You'll want to make piles of them for Asian-style hors d'oeuvre. Arrange them on skewers at a party as the centerpiece, and watch everybody grab for the last one.

Ingredients

1 cup club soda
1 cup rice flour
½ cup balsamic vinegar
1 cup port wine
2 Tbsp. wasabi powder
2 Tbsp. hot water
1 cup sour cream
15–20 medium-size morels
Vegetable oil

Preparation

TEMPURA BATTER

Whisk club soda and rice flour together and let stand 3 minutes.

BALSAMIC-PORT REDUCTION

In a stainless steel saucepan over medium-high heat, combine vinegar and port, reduce by half, and set aside.

WASABI CREAM

Mix wasabi powder and water, cover, and let stand 1 minute. Add sour cream, combine, and set aside.

MUSHROOMS

To have perfect-looking mushrooms, vegetable oil must be heated to 365 degrees. Carefully dip mushrooms in tempura and let excess batter drain off. Slowly add mushrooms to oil one at a time. Fry on each side 3 minutes and carefully transfer to a paper towel–lined plate. Assemble morels on one side of plate and spoon balsamic-port reduction on the other side. Add a few dollops of wasabi cream.

Hen-in-a-Basket

Lasse Sorensen, executive chef
TOM'S PLACE, DESOTO, ILLINOIS

The meaty texture of the mushroom known as Hen-of-the-Woods is big enough to star in this main course for mushroom lovers, or it could also be a hefty appetizer. This rich preparation features puff pastry baskets loaded with chunks of mushroom in a thick rosemary cream sauce. Since just one Hen-of-the-Woods can weigh 10 pounds or more, there's almost no limit to how many dinner guests you can invite. If you don't have Hen-of-the-Woods, Chicken-of-the-Woods (p. 38) is an easy choice to substitute. Experiment with other species to create your own medley of mushrooms in puff pastry.

Ingredients

- 2 cups Hen-of-the-Woods mushroom (p. 35)
- 3 Tbsp. butter
- 1 clove garlic, diced
- ¼ cup white wine
- ½ tsp. chopped fresh rosemary, reserving a sprig for garnish
- 1 cup cream or half-and-half
- Salt and pepper

Preparation

Roughly chop and break apart mushroom into bite-size pieces. In a large skillet over medium-high heat, melt butter until it stops foaming and clears, then add mushrooms and garlic. Cook for several minutes, stirring occasionally, until mushrooms turn golden brown and release their best flavor. Add wine, cook briefly until liquid is reduced by one-third, and then add chopped rosemary, mixing it well among the mushrooms. Add cream and continue simmering for a minute or two until sauce thickens, then ladle over store-bought puff pastry shells/baskets and serve with a fresh rosemary sprig for garnish. Serves four.

TIP: A tiny bit of rosemary goes a long way. Don't be tempted to add more than ½ tsp. to flavor this dish properly.

Morel Tiramisu

Chefs Gert and Lasse Sorensen
TOM'S PLACE, DESOTO, ILLINOIS

Morels marinated in Kahlua infuse this classic dessert with an inexplicably delightful mushroom flavor. This is one of those impossible-to-believe dishes that makes total strangers in restaurants offer a fork to passers-by and beg, "You absolutely must try this."

A *mushroom* dessert. Nobody ever sees it coming.

Ingredients

10 cups espresso (or extremely strong coffee, to which up to 5 Tbsp. instant coffee can be added to increase strength)
2 shots Kahlua
1 shot spiced Jamaican rum (such as Captain Morgan)
1 shot Bacardi rum
1 shot Frangelica
4 egg yolks
4 Tbsp. sugar
16 oz. mascarpone cheese
1 quart heavy cream
2 Tbsp. butter
20 morel mushroom caps, chopped
60 ladyfinger cookies
Cocoa powder

Preparation

Chill espresso or coffee. Mix 1 shot each of Kahlua, both varieties of rum, and Frangelica, then add half of the liquor mixture to chilled coffee. In a separate bowl, whip together egg yolks and sugar until foamy. In another bowl, add remaining half of liquor mixture to softened mascarpone cheese, whipping together to blend without lumps. Add the egg and sugar mixture to the cheese, blending well. In another bowl, whip the cream, then fold into mixture. In a large skillet over medium-

high heat, melt butter and sauté morels 3–5 minutes, until they begin to pop, then add the remaining shot of Kahlua and reduce, simmering until mushrooms appear caramelized. Dunk ladyfingers into coffee mix, soaking well, then arrange side by side on the bottom of a 10-inch-by-5-inch loaf pan. Spoon cheese mixture on top of cookies, creating another layer about the same depth as cookie layer. Add a third layer of caramelized morels. Repeat layers of the alternating ingredients, in the same order, until pan is nearly full. Sprinkle with cocoa powder and chill in refrigerator for at least 6 hours or overnight.

NOTE: For better appearance, lay cookies in different directions with each layer, as one might stack firewood in opposite-facing rows. European confectionery and pastry expert Gert Sorensen prepared the more complex version of this dessert seen in the photograph. Your results will look different.

Parasol Mushroom Frittata

Paul Virant, executive chef

VIE, WESTERN SPRINGS, ILLINOIS

The Parasol Mushroom has no match among gilled mushrooms when it comes to its complex, velvet-smooth flavor. But add farm-fresh eggs, Illinois goat cheese (Prairie Fruits Farm in Champaign, Illinois, makes it right), and bits of buttery Yukon Gold potatoes—now there's a real match. This frittata can be served warm as a breakfast entrée or chilled and eaten from a picnic basket while hunting for more Parasols during late summer and early autumn forays.

Ingredients

2 Tbsp. grape seed oil
1 small Yukon Gold potato, peeled and diced small
2–3 oz. Parasol Mushroom caps, sliced (6–8 caps, depending on size) (p. 150)
1 garlic clove, sliced
1 tsp. butter
2 Tbsp. fresh goat cheese
3 eggs, well beaten
Salt and pepper

Preparation

Turn on broiler and preheat an 8-inch nonstick pan or well-seasoned cast-iron skillet, add oil. Add diced potatoes and cook until golden brown, about 4 minutes. Add mushrooms and garlic, cook for another minute, add the butter. Add the goat cheese and eggs and season with salt and pepper. Stir constantly with a heat-resistant rubber spatula for 30 seconds to scramble eggs. Allow egg mixture to set on the bottom and cook for another minute. Place under broiler for about 2 minutes to cook the top. Invert on a serving plate. Cut into four wedges and serve with a small green salad. Serves four.

COMMENTS: Add enough Parasol Mushrooms to satisfy your preference. Too few mushrooms result in a very mild-tasting dish—still wonderful, but lacking the Parasol punch.

Black Trumpet–Crusted Walleye with Roasted Almonds and Lemon Vinaigrette

Paul Virant, executive chef
VIE, WESTERN SPRINGS, ILLINOIS

The lightest, most delicious freshwater fish in Illinois is the walleye. Its firm, white, flaky texture and mild flavor make it the top selection of both anglers and chefs. Pair it with the darkest, most delicious of Illinois mushrooms, the Black Trumpet, and you've got a perfect culinary contrast. A crust prepared from Black Trumpets and bread crumbs seals a flavorful walleye filet in this black-and-white creation that also features roasted almonds and a zesty lemon vinaigrette.

Walleye and Black Trumpets spend their lives in elusive obscurity, hidden from our view by absolute darkness; walleye shun daylight, creeping toward shore only during evening. Black Trumpets defy us by disappearing among forest leaves under somber camouflage. Fortunately, both delicacies can be found—but with some effort. A mere fork is all that's required here.

Ingredients

BLACK TRUMPET CRUST

¼ oz. dried Black Trumpets (p. 109)
½ cup bread crumbs
2 oz. melted butter
Salt

ALMOND AND BLACK TRUMPET MIXTURE

2 tsp. butter
1 oz. dried Black Trumpets, rehydrated (about 1 cup fresh)
2 Tbsp. roasted almonds (Marcona almonds work best)
Salt and pepper

LEMON VINAIGRETTE

1 shallot, minced
1 tsp. honey
1 lemon (juice and zest)
3 Tbsp. olive oil
Salt and pepper

WALLEYE

2 Tbsp. grape seed oil
Salt and pepper
4 2–½ oz. portions of walleye
Almond and Black Trumpet mixture
½ cup flat leaf parsley, chopped
Black Trumpet crust
Lemon Vinaigrette

Preparation

BLACK TRUMPET CRUST

Process Black Trumpets and bread crumbs in food processor until finely chopped. Add melted butter, mix, and reserve.

ALMOND AND BLACK TRUMPET MIXTURE

Heat the butter in a sauté pan, add the Black Trumpets and almonds, and cook for about 2 minutes. Season with salt and pepper and reserve.

LEMON VINAIGRETTE

Combine shallot, honey, and lemon juice and zest; whisk in olive oil, season with salt and pepper, and reserve.

ASSEMBLING THE WALLEYE

Turn on broiler and preheat a heavy-bottomed sauté pan with the grape seed oil to medium-high heat. Season walleye with salt and pepper and place bone-side down (do not move fish once in pan). Heat up the almond and Black Trumpet mixture, add the parsley to it, and keep warm. Cook fish for about 2 minutes until golden brown; remove from pan. Pack the Black Trumpet crust on the raw side of the fish; place fish back in broiler, crust side up, and cook for another 2 minutes until fish is done and the crust is crispy. Place almond and Black Trumpet mixture on four plates, put walleye on top, and garnish with lemon vinaigrette. Serves four as an appetizer.

· ·

Chanterelle Cream Soup

Thierry LeFeuvre, executive chef

FROGGY'S FRENCH CAFÉ, HIGHWOOD, ILLINOIS

There are many ways to prepare Yellow Chanterelles, but there's absolutely no way to prepare a Yellow Chanterelle soup that's more delicious than this luxuriously rich French offering. A hint of thyme and a touch of bay leaf define this exquisite, chicken-broth-based culinary treasure. When chanterelles are in season during mid-summer, you'll want to share the bounty. A summer dinner party calls for seasonal delicacies, and this dish celebrates the beloved Yellow Chanterelle with unforgettable richness. This recipe creates enough portions to satisfy more than a dozen mushroom lovers—with some to spare.

Ingredients

3 oz. olive oil
3 slices bacon (or smoke flavor)
2 Spanish onions, sliced
3 cloves garlic, chopped
1 bay leaf
1 tsp. thyme
¼ cup flour
1 gallon chicken broth
3 lbs. Button Mushrooms, washed and sliced
½ lb. Yellow Chanterelle mushrooms, sliced (p. 100)
3 shallots, chopped
2 garlic cloves, chopped
2 oz. dry white wine
8 oz. whipping cream
Chopped chives for garnish

Preparation

In a large saucepan, heat olive oil and cook bacon slices. Once bacon has browned, add the onion and garlic and cook until nicely browned. Add the thyme and bay leaf (adjust to taste). Add the flour and mix together. Add chicken broth and Button Mushrooms. In a separate sauté pan, sauté the chanterelles (reserve some for garnish) with the shallots and garlic. When nicely browned, add the white wine and cover for 2 minutes. Add the chanterelle mixture to the soup. Once the soup is brought to a boil, cover with a lid and simmer for approximately 45 minutes. When finished, add the cream and strain the soup. Soup can be served hot or cold, and topped with chives for garnish. Serves twelve.

Wild Mushroom Lasagna with Arugula Pesto

Charlie Trotter, executive chef

CHARLIE TROTTER'S, CHICAGO, ILLINOIS

Who says all lasagna must be made exactly the same? In this Italian classic, the mushrooms—used instead of tomatoes—really shine, with arugula pesto providing the perfect piquant accent. The result is a combination of flavors that is both earthy and heady. This is also a fine make-ahead dish: Just heat up the desired portion whenever you need it. For waterfowlers, add shredded duck confit between the layers for an even heartier filling.

Ingredients

ROASTED MUSHROOMS

3 cups assorted wild mushrooms, chopped (Yellow Chanterelles, Oyster Mushrooms, Black Trumpets, and/or Meadow Mushrooms)
½ cup Spanish onion, chopped
1 clove garlic, chopped
1 sprig thyme or rosemary
2 Tbsp. olive oil
¾ cup mushroom stock or water
Salt and pepper

LET'S EAT

MUSHROOM FILLING

5 cups roasted wild mushrooms (see preparation above), finely chopped
3 Tbsp. balsamic vinegar
3 Tbsp. extra virgin olive oil
3 Tbsp. fresh basil, chopped
Salt and freshly ground black pepper

RICOTTA FILLING

2 cups ricotta cheese
1 cup blanched spinach leaves, chopped
1 Tbsp. balsamic vinegar
1 clove garlic, minced
Salt and freshly ground black pepper

LASAGNA

1 lb. lasagna noodles, cooked
Mushroom filling
2 cups Pecorino Romano cheese, grated
1 lb. (2 large bunches) arugula leaves, washed
Ricotta filling

ARUGULA PESTO

2 cups firmly packed arugula leaves
1 ¼ cups firmly packed fresh basil
2 cloves garlic, chopped
1 cup grated Pecorino Romano cheese
1 cup extra virgin olive oil
1 ¼ cups pine nuts, toasted
Salt and freshly ground black pepper

GARNISHES

Arugula pesto
4 teaspoons balsamic vinegar
2 tbsp. basil cut into shreds
1 ½ cups pine nuts, toasted and chopped
2 tbsp. Pecorino Romano cheese, grated

Preparation

ROASTED MUSHROOMS

Place the mushrooms in an ovenproof pan and toss with the onion, garlic, thyme or rosemary, and olive oil. Add the stock or water and season with salt and pepper. Cover and bake at 325 degrees for 30 to 40 minutes, or just until the mushrooms are tender. Remove from the oven and cool in the cooking juices. Yields about 2 cups.

MUSHROOM FILLING

Combine roasted mushrooms, vinegar, olive oil, and basil. Season to taste with salt and pepper.

RICOTTA FILLING

Combine the ricotta, spinach, vinegar, and garlic. Season to taste with salt and pepper.

LASAGNA

Preheat oven to 375 degrees. Lightly oil an 8-inch-square lasagna pan. Arrange a layer of lasagna noodles on the bottom of the pan (trim the noodles as necessary to fit). Then begin layering by spreading a layer of the mushroom filling, sprinkling with some of the Pecorino Romano, adding a layer of arugula leaves, and spreading some of the ricotta filling over the arugula. Top with a second layer of the lasagna noodles. Continue this process, ultimately using half of the Pecorino Romano and all of the other components, to make 6 layers of pasta and 5 layers of filling. Cover with aluminum foil and bake for 45 minutes. Remove the foil and sprinkle with the remaining 1 cup Pecorino Romano. Continue to bake for 10 to 15 minutes, or until the cheese is melted and lightly golden brown. Remove the lasagna from the oven and let rest for 15 minutes before serving.

ARUGULA PESTO

Puree the arugula, basil, garlic, Pecorino Romano, olive oil, and pine nuts in a blender until smooth. Season to taste with salt and pepper. If the puree is too thick to spoon around the plate, thin it with a little water or more olive oil. The pesto can be used at room temperature, or warmed slightly just prior to use.

TO SERVE

Place a 4-inch-square piece of the lasagna in the center of each plate and spoon the arugula pesto and balsamic vinegar around the lasagna. Sprinkle with the basil, pine nuts, and cheese.

. . . .

Duck Confit and Morel Wellington with Vanilla Bean–Game Reduction

Christian Phernetton, executive chef
SILVERCREEK, URBANA, ILLINOIS

Ingredients

MOREL MUSHROOM FILLING

¼ cup butter
2 cloves garlic, thinly sliced
2 shallots, finely minced
½ Tbsp. thyme, chopped
¾ cup morel mushrooms (or any wild mushroom you may find), thinly sliced
¼ cup Marsala or Madeira wine
2 slices white bread, crumbled
Salt and freshly ground black pepper

VANILLA BEAN–GAME REDUCTION SAUCE

2 Tbsp. cottonseed oil
1 lb. duck bones, roasted
1 vanilla bean, split lengthwise
1 tomato, quartered
1 Spanish onion, chopped
1 garlic clove, chopped
1 celery stalk, chopped
1 carrot, chopped
2 bay leaves
6 black peppercorns
2 cups chicken stock
2 cups veal stock

DUCK CONFIT WELLINGTON

2 12-inch-square sheets of all-butter puff pastry
Morel mushroom filling
1 cup duck confit
2 eggs, beaten
Vanilla bean–game reduction sauce

GARNISH

Tiny lettuce leaves or fresh herbs
Extra virgin olive oil

Preparation

MOREL MUSHROOM MIXTURE

In a large sauté pan over medium-high heat, add the butter, garlic, shallots, and thyme. Cook until translucent, then add the mushrooms and turn the heat to high. Stir the mushrooms every 2 minutes until cooked through. Turn the heat to low and add the Marsala or Madeira and bread. Cook another 5 minutes, stirring constantly. Cool the mixture and chop until smooth. Season to taste with salt and pepper.

SAUCE

In a large pot over medium-high heat, add the cottonseed oil, duck bones, and the rest of the ingredients except for the chicken and veal stock. Cook for 8 minutes or until golden brown. Add the chicken and veal stock and reduce until the sauce has a thick consistency. Strain the sauce through a fine sieve and keep warm.

TO ASSEMBLE AND SERVE

Cut the puff pastry into 4 6-inch square pieces. Place the morel mushroom filling evenly in the center of each piece of pastry. Spread the duck confit evenly atop the mushroom mix. Fold the dough over the meat on the left and right sides. Brush the exposed puff pastry with the beaten eggs. Then roll the puff pastry from bottom to top, ending with the seal on the bottom. Place the pockets on a piece of wax paper, seal side down. Brush the tops with remaining egg wash and refrigerate. Preheat oven to 350 degrees. Place the cold pockets in the preheated oven and bake until golden brown, about 25–30 minutes. Spoon some warm sauce on a plate. Place the Wellington pocket atop the sauce and garnish with tiny lettuce or fresh herbs and a drizzle of olive oil. Serves four.

Wild Mushroom Risotto with Black Truffles
Rick Tramonto, executive chef
TRU, CHICAGO, ILLINOIS

The fabulous Black Truffle (*Tuber melanosporum*) is known to occur in Illinois—but only in the best kitchens, not in the wild. It's traditionally found in France, where tons of these "black diamonds" get exported annually after being excavated from beneath host trees. In the Midwest, there actually are several different species of truffle that have been documented. Unfortunately, their disappointingly simple flavor makes hunting for truffles in Illinois a mycological triviality.

But Illinois does have Black Trumpets, sometimes called the poor man's truffle because of the culinary trick of dicing Black Trumpets to make them appear to be Black Truffle shavings. This fragrant and flavorful risotto incorporates a generous portion of tasty Black Trumpets—plus a collection of mixed store-bought mushrooms—that serves as the background for the coup de grâce of Black Truffle slices. Yes, Black Truffles are expensive—but a plane ticket to France isn't cheap either. Commercially available mushrooms such as Portobella, Shiitake, and Button Mushrooms are great here, but wild-harvested Meadow Mushrooms (p. 139) or Yellow Chanterelles (p. 100) are a tempting substitution.

Ingredients

6 Tbsp. unsalted butter
1¼ cup Black Trumpet mushrooms, sliced (p. 109)
Kosher salt and freshly ground black pepper
1¼ cup Shiitake Mushrooms, sliced
1¼ cup small Button Mushrooms, sliced
1¼ cup Portobella Mushrooms, sliced
3 Tbsp. cold whipping cream
4 quarts vegetable stock
½ white onion, minced
garlic, minced
1 lb. Arborio rice
½ cup freshly grated Grana Padana Parmesan cheese
1 to 2 Tbsp. fresh Black Truffles, finely sliced (or chopped truffles in oil)
1 Tbsp. chives, finely chopped

Preparation

In a small skillet, melt 1 Tbsp. butter over medium-high heat. Add the Black Trumpet mushrooms and sauté until softened and fragrant, about 3 minutes. Add salt and pepper to taste and set aside. Repeat procedure for Shiitake, small Button, and Portobella Mushrooms individually and set all aside. With a whisk, whip the cream until stiff peaks form. Cover and refrigerate until ready to use. In a saucepan, bring the stock to a simmer. In a deep, heavy saucepan over medium heat, melt 1 Tbsp. butter. Sweat onion and garlic until translucent. Add the rice and stir until completely coated with butter. Add a ladleful of stock and cook, stirring, until all of the stock is absorbed by the rice. Add another ladleful of stock and continue cooking in the same manner, adding the stock as needed and stirring very frequently, until the rice is tender, about 25 minutes. Stir in the sautéed mushrooms, Grana Padana Parmesan cheese, and remaining 4 Tbsp. butter and mix to combine. Fold in the whipped cream and half of the Black Truffles. Taste and, if necessary, add salt and pepper. Plate and garnish with remaining truffles and chopped chives. Serve immediately. Serves six.

Index

A

Agaricus, 14, 137–46; Agaricus bisporus (Button Mushroom), 138; Agaricus californicus, xi; Agaricus campestris (Meadow Mushroom), 138–40, 144; Agaricus cupreobrunneus, 146; Agaricus porphyrocephalus, 146; Agaricus silvicola, 142
Alcohol Inky Cap (Coprinus atramentarius), 148
Amanita, 10, 11, 16, 19, 24, 26, 27, 30, 136, 138, 140, 142, 151, 152, 155, 156; Amanita bisporigera (Destroying Angel), 11, 16, 26, 136; Amanita thiersii (Thiers Amanita), 19, 140, 142; Amanita virosa (Destroying Angel), 26
Amatoxins, 24, 26, 27
American Parasol (Lepiota americana, Leucoagaricus americanus), 154–57
A-1 Mushroom Sauce, 181
Armillaria (Honey Mushroom), 67–70; Armillaria gallica, 69; Armillaria mellea, 67, 69, 70; Armillaria tabescens, 67, 69, 70
Auricularia auricula-judae (Wood Ear), 65, 66

B

Bart, Felicia, 33
Bear Lentinus (Lentinellus ursinus), 50
Bear's Head (Hericium americanum), 61, 62
beer, 3, 184
beer batter, 161, 176
Beer Batter Morels, 176, 177
Berkeley's Polypore (Bondarzewia berkeleyi), 37, 42, 43
Big Red (Gyromitra caroliniana), 88
bioluminescence, 15, 28, 103
Black and Bleu Morels, 174
Black Morel (Morchella elata), 80, 86, 89, 90, 92, 94, 164, 174, 175; comparison with Half-Free Morel (Morchella semilibera), 92, 94, 164
Black-Staining Polypore (Meripilus giganteus), 37
Black Trumpet (Craterellus cornucopiodes), 9, 106, 107, 109–14, 160, 163, 172, 173, 182, 195–97, 199, 203, 204
Black Trumpet-Crusted Walleye, 195–97
Black Trumpet Salad, 172
Bleichner, Loyce, 181

Blonde Morel (Morchella esculenta), 85
Bolete, vii, 116, 117, 119–28, 182
Boletus bicolor, 128
Boletus edulis (Cep, Porcini), 120
Bolognese, mushroom, 188
Bondarzewia berkeleyi (Berkely's Polypore), 37, 42, 43
Box Turtle, Eastern, 78
Brain Puffball, 135
Branson, Eugene Cunningham, 12
Brown Meadow Mushroom (Agaricus sp.), 145, 146
Bryant, Tammy, 96
Bushnell, Mr. and Mrs. David, 97
Button Mushroom (Agaricus bisporus), 101, 122, 138, 140, 141, 145, 146, 197, 204

C

Calvatia craniformis (Brain Puffball), vii, 129, 135
Calvatia cyathiformis (Purple-Spored Puffball), vii, 129, 134
Calvatia gigantea (Giant Puffball), vii, 129, 130, 134
Cantharellus (Chanterelles), vii, 5, 29, 99, 100, 102, 104, 106, 115, 162; Cantharellus cibarius (Yellow Chanterelle), vii, 99, 100, 102, 115, 162; Cantharellus cinnabarinus (Red Chanterelle), vii, 99, 106; Cantharellus lateritius (Smooth Chanterelle), 29, 99, 104
Cauliflower Mushroom (Sparassis hebstii), 63, 64
Cep (Porcini, Boletus edulis), 120
chanterelle, viii, xii, 5, 11, 28, 29, 71, 99–108; Red Chanterelle (Cantharellus cinnabarinus), 106–8; Smooth Chanterelle (Cantharellus lateritius,) 29, 104–6; Yellow Chanterelle (Cantharellus cibarius), 11, 28, 29, 100–107, 115–17, 161; recipes, 197, 199, 203

Chanterelle Cream Soup, 197
Chestnut Bolete (Gyroporus castaneus), viii, 119–22
Chicken and Lobster Sunrise, 166
Chicken Laccaria, 179
Chicken Mushroom (Laetiporus spp), 14, 28, 38–43; recipes, 160, 166, 167
Chicken-of-the-Woods (Chicken Mushroom, Laetiporus spp), 38, 39
Chicken Parmesan, 182
Chlorophyllum molybdites (Green-Spored Lepiota), 16, 21–23, 30, 138, 142, 151, 153
Clavulina zollingeri, 2
Conway, Lee, 168
Cool Oyster Salad, 184
coprine, 17, 20, 148
Coprinus atramentarius (Alcohol Ink Cap), 148
Coprinus comatus (Lawyer's Wig, Shaggy Mane), 13, 147–49; use of ink for writing, 148
Coprinus micaceus, 149
Coral Mushroom, 61, 62
Cortinarius, 57, 58, 74, 77
Craterellus cornucopioides (Black Trumpet), 9, 99, 100, 106, 107, 109–14, 160, 163; recipes, 172, 173, 182, 195–97, 199, 203, 204
Craterellus foetidus (Fragrant Black Trumpet), 99, 112
Crepidotus, 51
Crimini, 145
Cyclosporine, 7

D

Deadly Galerina (Galerina autumnalis group), 24, 25, 55, 70
deliquescence, 148, 149
Destroying Angel (Amanita bisporigera), 11, 26, 27; as puffball look-alike, 136
Devil's Urn (Urnula craterium), 111
Double Oyster Chowder, 170

Dryad's Saddle (Pheasant Back, Polyporus squamosus), 44
Duck Confit and Morel Wellington, 201

E

Elder, "Speedy," 96
Elephant Ears (Oyster Mushroom, Pleurotus ostreatus), 48
Elm, American, 44, 45, 86
Elm Oyster (Hypsizygus tessulatus), 48
Enchanted Grifola, 185
Exida, 66

F

False Morels (Gyromitra species), 88
Flammulina velutipes (Velvet Foot), 24, 25, 53–55
Fluted Black Elfin Saddle (Helvella laucunosa), 114
Fontana, Mark, 173
Fragrant Black Trumpet (Craterellus foetidus), 112–14
freezing morels, 164, 165

G

Galerina autumnalis (Deadly Galerina), 24, 25, 53, 55, 70
Galiella rufa, 111, 136
Giant Puffball (Calvatia gigantea), 130, 131, 134, 135; recipe, 184
gill color, as a diagnostic trait, 144
Gray Morel (Morchella esculenta), 85
Green-Spored Lepiota (Chlorophyllum molybdites), 16, 21–23, 30, 151
Grifola frondosa (Hen-of-the-Woods), 13, 35–37, 39; recipes, 185, 186, 191
Gyromitra (False Morels), 88; Gyromitra brunnea, 88; Gyromitra caroliniana, 88
Gyroporus castaneus (Chestnut Bolete), viii, 119–22

H

Hagler, Justin, 179
Half-Free Morel (Morchella semi-libera), 80, 105; comparison with False Morels, 88, 91, 92; comparison with Verpa conica, 84
Hapalopilus croceus, 42
Hedgehog Mushroom (Hydnum repandum), 115–17
Helvella lacunosa, 114
Hen-in-a-Basket (recipe), 191
Hen-of-the-Woods (Grifola frondosa), 35–37, 39; recipes, 185, 191
herbicide poisoning, 138
Hericium americanum (Bear's Head), 13, 61, 62
Hericium coralloides, 61
Hericium erinaceus (Lion's Mane), 59–62; recipe, 177; subspecies erinaceo-abietis, 60
Honey Mushroom (Armillaria species), 24, 25, 67–70; bioluminescence of mycelium, 103
Humungous Fungus, 67
Hydnum repandum (Hedgehog Mushroom), 115, 116
Hygrocybe coccinea (Scarlet Waxy Cap), 3, 108
Hygrophoraceae (Waxy Caps), 108
Hygrophorus pratensis (chanterelle look-alike), 102
Hypomyces lactifluorum (Lobster Mushroom), 71, 72
Hypsizygus tessulatus (Elm Oyster Mushroom), 48

I

ibotenic acid, 20
Illinois Mycological Association, 8
Illinois State Morel Hunting Championship, 81
Indigo Milk Cap (Lactarius indigo), xi, 75–77; quality of flavor, 161
ink, from Coprinus comatus, 148
inky cap, 147–49
Inonotus, 42

J

Jack O'Lantern (Omphalotus illudens), 28, 29; as chanterelle look-alike, 103
Judas' Ear (Wood Ear, Auricularia auricula-judae), 65, 66

K

Kerrigan, Richard, xiii, 146

L

Laccaria ochropurpurea (Purple-Gilled Laccaria), 57, 73, 74; recipes, 173, 179
Lactarius, 71; Lactarius indigo (Indigo Milk Cap), xi, 75–77
Laetiporus cincinnatus (Chicken Mushroom with white underside), 38
Laetiporus sulphureus (Chicken Mushroom with yellow underside), 13
Lasagne, Wild Mushroom, 199
Lawyer's Wig (Coprinus comatus, Shaggy Mane), 13, 147–49
LBM (little brown mushrooms), 25
Lee, Bao Cheng, 184, 185, 187
Lentinellus ursinus (Bear Lentinus), 50
Lepiota, Green-Spored (Chlorophyllum molybdites), 16, 21–23, 30, 138, 142, 150–53
Lepiota americana (American Parasol, Leucoagaricus americanus), 154–57
Lepista nuda (Wood Blewit), 56–58
Leucoagaricus americanus (American Parasol, Lepiota americana), 154–57
Leucoagaricus naucinus, 143
Lilac Bolete (Xanthaconium separans), 119, 126–28
Lion's Mane (Hericium erinaceus), 59; recipe, 178
Lobster Mushroom (Hypomyces lactifluorum), 71, 72
Lycoperdon perlatum, 13, 133
Lycoperdon pyriforme (Pear-Shaped Puffball), viii, 129, 132, 133

M

Macrolepiota procera (Parasol Mushroom), 21–23, 150–53, 155; cooking with, 160; recipe, 194
Macrolepiota rachodes (Shaggy Parasol), 22
Marasmius pyrocephalus, 55
McGovern, Andrew, 176, 189
Meadow Mushroom (Agaricus campestris), 10, 16; Agaricus, 138–46; poisonous look-alike, 30; recipes, 182, 199, 203
medicinal mushrooms, 17
Meripilus giganteus (Black-Staining Polypore), 37
Mica Cap (Coprinus micaceus), 149
Missouri Mycological Society, 8
monomethyhydrazine (MMH), 88
Morchella (morels), viii, 79–98; Morchella elata (Black Morel), viii, 79, 80, 89, 164; Morchella esculenta (Yellow Morel), viii, 79, 80, 84, 85, 93; Morchella semilibera (Half-Free Morel), 79, 80, 84
morels, viii, 8, 11, 12, 44, 45, 79–98, 111; flavor of, 161, 162; freezing, 164, 165; how to cook, 170; medicinal use, 17; as powder, 163; recipes, 168, 174, 176, 189, 192, 201. *See also* individual types
Morel Tiramisu, 192
The Mushroom Book, 12
mushroom flavors: about, 161–63
mushroom poisoning, 20, 144
mushroom powders: about, 163
mushrooms: how to cook, 160
Mushrooms over Toast, 181
Mushrooms with Bacon, 181

Mutinus elegans (Dog Stinkhorn), 87
mycelium, 4
Mycena leaiana, 108
mycophagists, wild, 78
mycorrhiza, 4, 6, 7, 34

N

Nauman, Tom and Vicky, 81
Nelson, Jim and Shirley, 96
North American Mycological Association, 8

O

Old Man of the Woods (Strobilomyces spp), 123
Omphalotus illudens (Jack O'Lantern), 28
Oyster Mushroom (Pleurotus ostreatus), 14, 24; recipes, 170, 184, 185, 187; Mock Orange (Phyllotopsis nidulans), 50

P

Panellus, 51
Parasol Mushroom (Macrolepiota procera), 23, 150, 160, 161
Parasol Mushroom Frittata, 194
Pear-Shaped Puffball (Lycoperdon pyriforme), 132
Phallus hadriani (Morel Stinkhorn), 87
Phallus impudicus (stinkhorn), 87
Pheasant Back (Dryad's Saddle, Polyporus squamosus), 44
Phernetton, Christian, 201
Pholiota, 68
Phyllotopsis nidulans (Orange Mock Oyster Mushroom), 50
Pleurotus ostreatus (Oyster Mushroom), 46
Pluteus, 49
poisoning, caused by herbicide, 138
poisonous mushrooms, 19
Polyporus squamosus (Pheasant Back, Dryad's Saddle), 44
Porcini (Cep, Boletus edulis), 120
Portobella, 145
preservation, semi-dry, 164
puffball, 26, 129
Puffball S'mores, 183
Purple-Gilled Laccaria (Laccaria ochropurpurea), 73
Purple-Spored Puffball (Lycoperdon cyathiformis), 134

R

recipes, 159
Red Chanterelle (Cantharellus cinnabarinus), 106–8
rhizomorph, 5
River Red (Gyromitra caroliniana), 88
Russula, 15; Russula brevipes, 71

S

Sagan, Carl, 15
Scallop-Hericium Checkerboard, 177
Scleroderma citrinum, 136
semi-dry preservation (freezing mushrooms), 164, 165
Shaggy Mane (Coprinus comatus), 147
Shaggy Parasol (Macrolepiota rachodes), 22
Shiitake, 159
Smith, Jerry, 97
Smooth Chanterelle (Cantharellus lateritius), 29, 104–6
Sorensen, Gert, 192, 193
Sorensen, Lasse, 166, 170, 192
Sparassis crispa (Cauliflower Mushroom, western United States), 63
Sparassis herbstii (Cauliflower Mushroom), 63
spore print: how to make, 14, 16
Steak Mushrooms, 173
stinkhorn, 87

Strobilomyces (Old Man of the Woods), 119; Strobilomyces confusus, 123; Strobilomyces floccupus, 123
Stropharia, 144
Sulphur Shelf (Chicken Mushroom, Laetiporus spp), 38

T

Tempura Morels, 189
Tolypocladium niveum, 7
Tremella foliacea, 66
Tremella mesenterica, 66
Tremellodendron pallidum, 64
Trotter, Charlie, 199
truffles, 161, 203

U

universal veil, 27
Urnula craterium (Devil's Urn), 111

V

Velvet Foot (Flammulina velutipes), 24
Verpa, 84; Verpa conica, 84
Virant, Paul, 194

Volk, Tom, 7
Volvariella, 49

W

Waxy caps (Hygrophoracea), 108
Wild Mushroom Risotto with Black Truffles, 203
Wild Turkey and Morels, 168
Witch's Butter (Exida sp.), 66
Wood Blewit (Lepista nuda), 56
Wood Ear (Auricularia auricula-judae), 65

X

Xanthaconium separans (Lilac Bolete), 119
Xerula radicata, 7

Y

Yellow Chanterelles (Cantharellus cibarius), 11, 28, 29, 100–107, 115–17, 161; recipes 197, 199, 203
Yellow Morel (Morchella esculenta), 16, 17, 80, 84–86; comparison with Black Morel, 89, 90; flavor of, 162; hunting tips, 93–95, 97

JOE McFARLAND is a nature writer and photographer whose stories and photographs have been published in national magazines, newspapers, and books. For eight years he was a syndicated outdoor writer for dozens of Illinois newspapers, including the *Springfield State Journal-Register*. Since 2002, he has been staff writer and a photographer for the Illinois Department of Natural Resources magazine *Outdoor Illinois*. McFarland's interest in mycology began more than thirty years ago and has since developed into a professional and culinary passion. His mushroom photographs have appeared on television, in newspapers, magazines, and calendars. He is an active member of regional and national mycological organizations, including the North American Mycological Association, the Mycological Society of America, the Illinois Mycological Association, and the Missouri Mycological Society. He has contributed fungi articles to magazines, newspapers, and journals, and teaches a course on morel mushrooms at a southern Illinois college with mycologist Walter J. Sundberg. He has appeared on television and lectures widely on mushrooms and other fungi at seminars and events including culinary classes at Kendall College in Chicago.

DR. GREGORY M. MUELLER received his BA and MS degrees from Southern Illinois University–Carbondale and his PhD from the University of Tennessee. He undertook postdoctoral work at Uppsala University in Sweden and at the University of Washington in Seattle before joining the staff at Field Museum of Natural History in 1985 as curator of fungi in the Department of Botany. In January 2009, Greg joined the Chicago Botanic Garden as Vice President, Science and Academic Programs. He is an adjunct faculty member at University of Chicago, University of Illinois at Chicago, and Northwestern University. In addition to working on northern Illinois and northwest Indiana fungi, Greg has carried out extensive fieldwork throughout Central and South America, parts of Europe, China, Australia, New Zealand, and Papua New Guinea. Greg's research and teaching activities focus on the diversity, distribution, ecology, and conservation of macrofungi. Greg serves as the scientific advisor for the Illinois Mycological Association and is "on call" for the Illinois Poison Control Center. He is a former President of the Mycological Society of America. Greg has published or edited five books on fungi as well as over one hundred articles in professional and popular journals.

The University of Illinois Press

is a founding member of the

Association of American University Presses.

Composed in 10.5/13 Adobe Minion Pro

with Meta display by Jim Proefrock

at the University of Illinois Press

Designed by Copenhaver Cumpston

Printed in Korea by Tara Printing

through Four Colour Book Group

University of Illinois Press

1325 South Oak Street

Champaign, IL 61820-6903

www.press.uillinois.edu